T0257883

Handbook of Environmental Monitoring

Handbook of Environmental Monitoring

Edited by **Emma Layer**

New York

Published by Callisto Reference,
106 Park Avenue, Suite 200,
New York, NY 10016, USA
www.callistoreference.com

Handbook of Environmental Monitoring
Edited by Emma Layer

International Standard Book Number: 978-1-63239-391-3 (Hardback)

Printed in the United States of America.

Contents

Preface

This book includes contributions by renowned scientists and researchers from across the globe. It presents current research advances and developments in the area of environmental monitoring to a worldwide audience of technicians, scientists, environmental educators, administrators, managers, students and those interested in the environment. The book discusses topics such as usage of wireless sensor networks or geo-sensor webs in the field of environmental monitoring.

This book is a comprehensive compilation of works of different researchers from varied parts of the world. It includes valuable experiences of the researchers with the sole objective of providing the readers (learners) with a proper knowledge of the concerned field. This book will be beneficial in evoking inspiration and enhancing the knowledge of the interested readers.

In the end, I would like to extend my heartiest thanks to the authors who worked with great determination on their chapters. I also appreciate the publisher's support in the course of the book. I would also like to deeply acknowledge my family who stood by me as a source of inspiration during the project.

Editor

Environmental Monitoring with Wireless Sensor Network Technology

Biosensor Arrays for Environmental Monitoring

Wei Song[1], Si Wei[2], Hong-Xia Yu[2], Maika Vuki[3] and Danke Xu[1]
[1]State Key Laboratory of Analytical Chemistry for Life Science,
School of Chemistry and Chemical Engineering, Nanjing University,
[2]State Key Laboratory of Pollution Control and Resource Reuse,
School of the Environment, Nanjing University,
[3]College of Natural and Applied Sciences, University of Guam, Mangilao, Guam,
[1,2]China
[3]USA

1. Introduction

Environmental monitoring involves several steps such as sampling, sample handling and sample transportation to specialized laboratories, sample preparation and analysis. Traditional environmental monitoring approaches are based on discrete sampling methods followed by laboratory analysis. These approaches do not improve our understanding of the natural processes governing chemical species behavior, their transport and bioavailability, or the relationship between anthropogenic releases and their long-term impact on aquatic systems[1]. The challenge of environmental monitoring in situ requires new and improved analytical devices featuring precision, sensitivity, specificity, rapidity, and ease of operation to detect decreasing concentrations of an ever growing array of pollutants. Such devices must be comparable to or better than traditional analytical systems, and must be simple to handle, small, cheap, able to provide reliable information in real-time, and must be sensitive and selective for the analyte of interest, and suitable for in situ monitoring[2]. Biosensors not only fulfill all these requirements but also have applications in many areas such as clinical diagnostics, forensic chemistry, pharmaceutical studies, food quality control and environmental monitoring.

A biosensor is an analytical device for the detection of an analyte that combines a biological component with a physicochemical detector component. It consists of 3 parts: (1) the sensitive biological element (biological material such as tissue, microorganisms, organelles, cell receptors, enzymes, antibodies, nucleic acids, etc.), a biologically derived material or biomimic; (2) the transducer or the detector element (works in a physicochemical way; optical, piezoelectric, electrochemical, etc.) that transforms the signal resulting from the interaction of the analyte with the biological element into another signal that can be more easily measured and quantified; (3) associated electronics or signal processors that are primarily responsible for the display of the results in a user-friendly way[3]. Depending on the type of transduction mechanism applied and the bio-recognition element employed, the potential for these devices for detection can be enormous. The technological development and the success in single analyte detection propelled advances in the miniaturization of sensors along with multi-analyte detection with sensitivities ranging in the nano-mole to atto-mole range. With advances in techniques for biosensor construction, it has been possible to miniaturize the whole biosensor system on a chip to fabricate biosensor arrays.

The Biosensor arrays developed at the Naval Research Laboratory (NRL) has successfully been used in the detection of a variety of protein toxins, organic molecules, physiological health markers, a virus and a number of bacteria, initially in buffer but increasingly in food, biological and environmental matrices [4]. These developed biosensors are rapid, simple to perform and require little-to-no sample pretreatment prior to analysis, even for more complex sample matrices. In addition, the two-dimensional nature of the slide sensing surface facilitates simultaneous analysis of multiple samples for multiple analytes. Research on biosensor arrays as multi-analyte bio-systems has generated increased interest in the last decade. The main feature of the micro-array technology is the ability to simultaneously detect multiple analytes in one sample by an affinity-binding event at a surface interface. Fifteen years ago, the gene expression analysis of cDNA on micro-arrays was one of the first applications that successfully detected thousands of labeled target DNA molecules in parallel. Also the first immuno-analytical biosensor array was described at the same time. In the meantime, a great variety of target analytes capable of interacting selectively with a bio-molecular receptor has been adapted to arrays[5]. The biosensor arrays have been envisioned as a tool for rapid, on-site screening of pollutants in whatever location they might be found. The goals of automation, weight reduction, minimal size, ease of use, and reliability have remained paramount as the system has been developed[6].

The challenge of continuous in situ monitoring of environmental pollution requires instruments that are robust and with sufficient sensitivity and long lifetime. Commonly used conventional methods are time-consuming, expensive, require skilled operators, and lack the required selectivity. Biosensor arrays have the advantage of being simple, uniform whole structures featuring direct transduction, high bio-selectivity, high sensitivity, miniaturization, electrical/optoelectronic readout, continuous monitoring, ease of use, and cost effectiveness. User advantages include low price, reliability, no sample preparation, disposability, and clean technology. Hence, biosensor arrays show the potential to complement both laboratory-based and field analytical methods for environmental monitoring. Biosensor arrays are based on one general principle—certain bio-molecular recognition elements are defined on a heterogeneous matrix. Each element is dedicated to an analyte and contains quantitative information. The matrix is a patterned surface where the recognition molecules are immobilized by micro-printing such as screen printed technique, micro fluidic or other micro-structuring processes[5]. The type of biosensor arrays involves DNA-based biosensor array, antibody-based biosensor array, aptamer-based biosensor array, enzyme-based biosensor array, and microorganism-based biosensor. Recent progress in the development of analytical detection methods for antibody arrays, enzyme arrays and aptamer arrays as well as microbial arrays are summarized in this review, and their applications in the environment monitoring are also discussed. Detection approach is focused on electrochemical and optical measurements including various electrochemical or florescent probes as well as label-free approach. The numerous fabrication methods of DNA capture probes, antibodies and aptamer for multiplexed biological targets are also discussed.

2. DNA-based biosensor arrays

Deoxyribonucleic acids (DNA) are arguably the most important of all bio-molecules. The unique complementary structure of DNA between the base pairs adenine/thymine and cytosine/guanine has been the basis for genetic analysis over the last few decades. The ability of a single stranded DNA (ssDNA) molecule to 'seek out', or hybridize to, its complementary strand in a sample is the foundation of DNA-based detection systems. There is a great potential market for simple, cheap, rapid, and quantitative detection of specific genes. Areas of application include clinical, veterinary, medico-legal, environmental, and the food industry[7]. Development of DNA biosensors and DNA biosensor arrays has increased tremendously over the past few years as demonstrated by the large number of scientific publications. Numerous DNA detection systems based on the hybridization between a DNA target and its complementary probe, which is present either in solution or on a solid support, have been described[8]. Homogeneous assays allowing the determination of DNA sequences have been developed. These systems can be based on optical[9] or electrochemical[10] detection. However, they do not allow easy continuous monitoring and miniaturization. Heterogeneous DNA biosensors and DNA biosensor arrays offer promising alternatives to these methods. They allow continuous, fast, sensitive, and selective detection of DNA hybridization, and they also can be reused. DNA biosensors arrays (commonly called gene chips, DNA chips, or biochips) exploit the preferential binding of complementary single-stranded nucleic acid sequences. This system usually relies on the immobilization of a single-stranded DNA (ssDNA) probe onto a surface to recognize its complementary DNA target sequence by hybridization. Transduction of hybridization of DNA can be measured optically, electrochemically, or using other devices. The detection process is schematized in Figure 1[8].

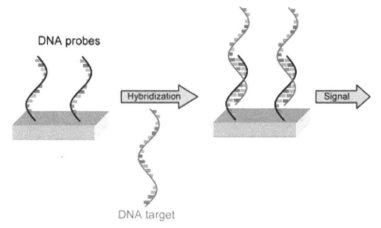

Fig. 1. Steps involved in the detection of a DNA sequence. Reprinted from ref. 8 with permission by the American Chemical Society.

In the case of DNA biosensors arrays, the immobilization of a DNA probe is achieved directly onto a transducer surface. DNA biosensor arrays are made from glass, plastic, or silicon supports and are constituted of tens to thousands of 10 – 100 µm reaction zones onto which individual oligonucleotide sequences have been immobilized. The exact number of DNA probes varies in accordance with the application. DNA biosensor arrays allow multiple parallel detection and analysis of the patterns of expression of thousands of genes in a single experiment.

We have presented an ultra-sensitive and direct electrochemical DNA biosensor array based on Ag aggregate tag and differential pulse voltammtery[11]. The scheme of detection is shown in Figure 2. The silver tags consist of Conjugate 1 (functionalized with capture probes and oligo A and Conjugate 2 (modified with oligo T). Hybridization between complementary oligo (d) A and oligo (d) T anchored on the silver nanoparticles produced aggregate tags. The hybridization-induced tags are successfully applied to bind with the DNA target via sandwich hybridization format and offer direct and amplified readout by differential pulse voltammetric method. We have found that the detection sensitivity by use of the aggregate tags can be improved by 3 orders of magnitude as compared to the single silver nanoparticle labels and a detection limit of 5 amol/L could be obtained.

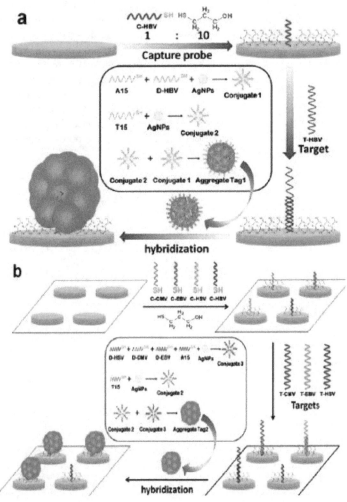

Fig. 2. Schematic illustration of the electrochemical Assay (a) and Multiplexed Assay (b) with silver nanoparticle Conjugates; Preparation of the aggregates is shown as well. Reprinted from ref. 11 with permission by the American Chemical Society.

Environmental applications of DNA biosensor arrays are in the field of species identification. For instance, DNA biosensor arrays are extensively exploited in the detection of pathogenic microorganisms relevant to food, bio-defense and environmental contamination applications. Mainly, DNA biosensor arrays have been coupled to PCR, as a specific detection method of the amplified base sequence. Zhang et al.[12] have developed a label-free electrochemical DNA biosensor array as a model system for simultaneous detection of multiplexed DNAs using micro-liters of sample. A novel multi-electrode array was comprised of six gold working electrodes and a gold auxiliary electrode, which were fabricated by gold sputtering technology, and a printed Ag/AgCl reference electrode was fabricated by screen-printing technology. The DNA biosensor array for simultaneous detection of the human immunodeficiency virus (HIV) oligonucleotide sequences, HIV-1 and HIV-2, was fabricated in sequence by self-assembling each of two kinds of thiolated hairpin-DNA probes onto the surfaces of the corresponding three working electrodes, respectively. The hybridization events were monitored by square wave voltammetry using methylene blue (MB) as a hybridization redox indicator. The oxidation currents of MB accumulated on the array decreased with increasing the concentration of HIVs due to higher affinity of MB for single strand rather than double strands of DNA. Under the optimized conditions, the peak currents were linear over ranges from 20 to 100 nmol/L for HIV-1 and HIV-2, with the same detection limits of 0.1 nmol/L (S/N= 3), respectively. The detection process is illustrated in Figure 3. The biosensor array showed a good specificity without the obvious cross-interference. Furthermore, single-base mutation oligonucleotides and random oligonucleotides can be easily discriminated from complementary target DNAs. Their work demonstrates that different hairpin-DNA probes can be used to design the label-free electrochemical biosensor array for simultaneous detection of multiplexed DNA sequences for various applications.

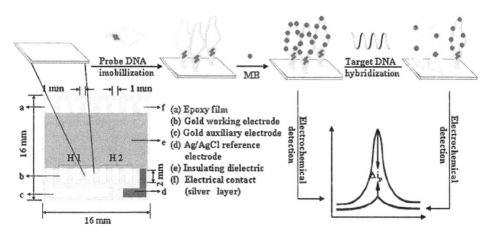

Fig. 3. Schematic diagrams of multi-electrode array and representation of biosensor array with fabrication steps and performance. Reprinted from ref. 12 with permission by the Elsevier.

Using electrochemical impedance spectroscopy (EIS) for biosensing applications typically requires repetitive experiments. To address this need, Bogomolova et al.[13] have designed a multi-specific electrochemical array with eight individually addressable 2 mm-diameter gold working electrodes for rapid biosensing data accumulation by EIS in the presence of redox agent. The array allows to incorporate multiple negative controls in the course of a single binding experiment, as well as to perform parallel identical experiments to improve reliability of detection. The array is fitted with attached electrochemical cell with Ag/AgCl mini reference electrode and can be used to process macro samples of 0.5–1 ml (Figure 4). The reported array is disposable, economical and is easy to use. Examples of array use for label-free genetic sensing of 2.7 kb-long target *Yersinia pestis* DNA and for protein sensing of Ricin Toxin Chain A (RTA) are presented. The authors suggest the reported array design as a tool for researchers in the area of EIS sensing.

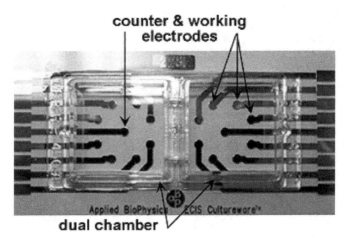

Fig. 4. Custom gold-sputtered dual array electrodes with attached chambers. Reprinted from ref. 13 with permission by the Elsevier.

Fu et al.[14] developed a piezoelectric quartz crystal microbalance (QCM) nucleic acid biosensor array using Au nanoparticle signal amplification to rapidly detect *S. epidermidis* in clinical samples. The synthesized thiolated probes specific targeting *S. epidermidis* 16S *r*RNA gene was immobilized on the surface of QCM nucleic acid biosensor arrays. Hybridization was induced by exposing the immobilized probes to the PCR amplified fragments of *S. epidermidis*, resulting in a mass change and a consequent frequency shift of the QCM biosensor. The results showed that the lowest detection limit of current QCM system was 1.3×10^3 CFU/mL. A linear correlation was found when the concentration of *S. epidermidis* varied from 1.3×10^3 to 1.3×10^7 CFU/mL. In addition, 55 clinical samples were detected with both current QCM biosensor system and conventional clinical microbiological method, and the sensitivity and specificity of current QCM biosensor system were 97.14% and 100%, respectively.

Doong et al.[15] fabricated a sol–gel-derived array DNA biosensors coupled with a fluorescence detection system and a robotic pin-printing platform to detect polycyclic aromatic hydrocarbons (PAHs) in water and serum samples (Figure 5). Parameters including sol percentage, doped-amount of glycerol, dye probes, the surface coating, and DNA concentration were optimized. In their work, two fluorescent dyes, fluorescein isothiocyanate (FITC) and ethidium bromide (EDB) were selected and compared. Results showed that EDB was more sensitive to compete intercalators (PAHs) than FITC, and was selected as fluorescent dye for array-based DNA biosensors. The optimized procedure with *ds*DNA concentration of 23.5 g/mL allowed the fabrication of the DNA biosensor up to 50 spots within 10 min via the developed pin-printing system. For PAH detection, the developed array DNA biosensor effectively detected naphthalene and phenanthrene in the concentration range of 0-10 mg/L in aqueous solution, but was not sensitive to fluoranthene and benzo[a]pyrene. In the serum samples, the apparent water solubility of high-molecular-weight PAHs was greatly enhanced by the dissolved organic compounds in serum, and an obvious DNA toxicity was exhibited in the presence of three-to-five-ring PAHs. Benzo[a]pyrene showed high toxic effect at low concentration in serum samples, clearly showing that the sol-gel-derived array DNA biosensor with EDB as sensing probe can effectively detect PAHs in water and biological samples.

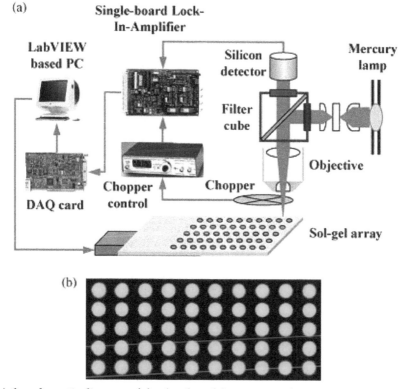

Fig. 5. (a) the schematic diagram of the developed fluorescence detection system and (b) images of the array DNA biosensor using the sol-gel processes. Reprinted from ref. 15 with permission by the Elsevier.

3. Antibody based biosensor array

Immunoassays gained popularity for biomedical applications in the 1970s because of the impressively low detection limits and high selectivity for analyzing complex samples that could be achieved with relatively simple procedures and instrumentation. The availability of highly selective antibodies for an increasingly wide variety of important analytes was also an important factor in the growth of the method over the following decades. The development of more sensitive labels and detection devices also improved the sensitivity of the assays even further. Once immunoassays became more common, the development of more convenient immuno-sensors that are easier and faster to use gained momentum[16].

Antibody-based biosensors are inherently more versatile than enzyme-based biosensors in that antibodies have been generated which specifically bind to individual compounds or groups of structurally related compounds with a wide range of affinities. There are, however, several limitations in the use of antibody-based biosensors for environmental monitoring applications. These limitations include the complexity of assay formats and the number of specialized reagents (e.g., antibodies, antigens, tracers, etc.) that must be developed and characterized for each compound and the limited number of compounds typically determined in an individual assay as compared to the multiple compounds that contaminate environmental samples[17].

Antibody-based biosensor arrays are a powerful tool for analytical purposes. Immuno-analytical micro-arrays are a quantitative analytical technique using antibodies as highly specific biological recognition elements[5]. They can be designed for a variety of analytical applications producing rapid results with low limits of detection (LOD). The detection antibodies are in direct contact with the sample, without prior sample cleanup. Immobilized haptens in combination with an indirect competitive immunoassay (IA) are the common format for the detection of multiple small molecules (e.g. pesticides, pharmaceuticals, small toxin targets). Haptens provoke an immune response if coupled to a protein by use of their functional groups. Hapten micro-arrays use analyte derivatives as immobilized recognition molecules. Antibody micro-arrays quantify proteins, bacteria, or viruses by using a sandwich immunoassay format. Recent advances reported for antibody-based biosensor arrays for environmental applications have primarily been focused toward these. For example, Jin et al.[18] have developed a fluorescent NP-mediated Ab micro-array system for the detection of bioterrorism agents exemplified by ricin, CT, and SEB toxins (results are displayed in Figure 6). High sensitivity, specificity, and reproducibility were achieved by using their antibody biosensor array. They found that substituting monoclonal antibodies (mAb) with highly purified polyclone antibodies (pAb), even though having similar titer by ELISA, could dramatically improve the micro-array performance. A likely explanation is that pAb enhances the Ag capture by binding multiple sites on the analyte. The micro-array format allows a multiplexed, high-throughput, and high-parallel assay. Furthermore, the miniature feature permits low consumption of both sample and reagents, reducing the amounts of biohazard waste as well as the costs to conduct the assay.

Seidel et al.[19] fabricated an automated chemiluminescence (CL) read-out system for analytical flow-through micro-arrays based on multiplexed immunoassays. The micro-array chip reader (MCR 3) is designed as a stand-alone platform, with the goal to quantify multiple analytes in complex matrices of food and liquid samples for field analysis or for routine analytical laboratories. The CL micro-array platform is a self-contained system for the fully automated multiplexed immuno-analysis comprising the micro-array chip, the fluidic system and the software module that enable automated calibration and determination of analyte concentrations during a whole working day. The detection of antibiotics in milk was demonstrated to validate this device. Therefore, an automated multi-analyte detection instrument is needed for the simultaneous and rapid quantification of antibiotics. Also regeneration is required to avoid replacing the assay surface. The European Union, for example, has defined maximum residue levels (MRLs) for a number of antibacterial compounds. However, despite the obvious demand for quantitative multi-residue detection methods that can be carried out on a routine basis, there is currently a lack in the development of such systems. In particular, an automated multi-analyte detection instrument is needed that is capable of quantifying several antibiotics simultaneously within minutes. Seidel's group[20] developed a new hapten based micro-arrays for the parallel analysis of 13 different antibiotics in milk within six minutes by applying an indirect competitive chemiluminescence micro-array immunoassay (CL-MIA). To allow multiple analyses, a regenerable micro-array chip was developed based on epoxy-activated PEG chip surfaces, onto which micro-spotted antibiotic derivatives like sulfonamides, b-lactams, aminoglycosides, fluorquinolones and polyketides are coupled directly without further use of linking agents. Using the chip reader platform MCR 3 (Figure 7), this antigen solid phase is stable for at least 50 consecutive analyses.

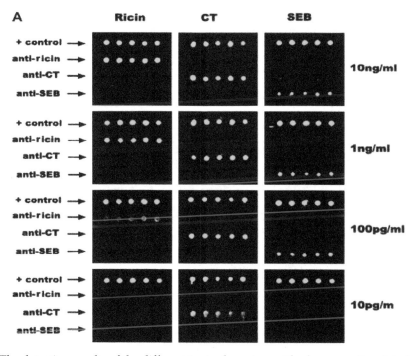

Fig. 6. The detection results of the different toxins by using antibody arrays. Reprinted from ref. 18 with permission by the Elsevier.

An impedance biosensor based on interdigitated array microelectrode (IDAM) coupled with magnetic nanoparticle–antibody conjugates (MNAC) was developed and evaluated for rapid and specific detection of E. coli O157:H7 in ground beef samples by Li et al.[21]. MNAC were prepared by immobilizing biotin-labeled polyclonal goat anti-E. coli antibodies onto streptavidin-coated magnetic nanoparticles, which were used to separate and concentrate E. coli O157:H7 from ground beef samples. Magnitude of impedance and phase angle were measured in a frequency range of 10 Hz to 1 MHz in the presence of 0.1mol/L mannitol solution. The lowest detection limits of this biosensor for detection of E. coli O157:H7 in pure culture and ground beef samples were 7.4×10^4 and 8.0×10^5 CFU/mL, respectively. The regression equation for the normalized impedance change (NIC) versus E. coli O157:H7 concentration (N) in ground beef samples was NIC=15.55N−71.04 with R^2= 0.95. Sensitivity of the impedance biosensor was improved by 35% by concentrating bacterial cells attached to MNAC in the active layer of IDAM above the surface of electrodes with the help of a magnetic field. Based on equivalent circuit analysis, it was observed that bulk resistance and double layer capacitance were responsible for the impedance change caused by the presence of E. coli O157:H7 on the surface of IDAM. Surface immobilization techniques, redox probes, or sample incubation were not used in this impedance biosensor. The total detection time from sampling to measurement was 35 min.

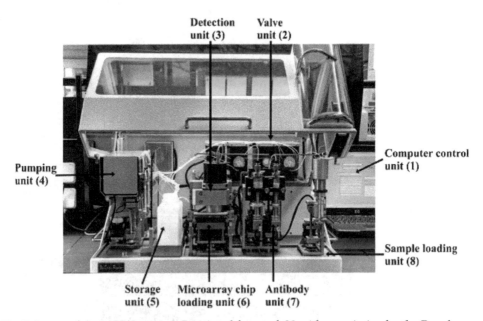

Fig. 7. Image of the MCR 3 system. Reprinted from ref. 20 with permission by the Royal Society of Chemistry.

Because of the potential health risks of aflatoxin B1 (AFB1), it is essential to monitor the level of this mycotoxin in a variety of foods. An indirect competitive immunoassay has been developed using the NRL array biosensor (Figure 8) by Shriver-Lake et al.[22], offering rapid, sensitive detection, and quantification of AFB1 in buffer, corn and nut products. AFB1-

spiked foods were extracted with methanol and Cy5-anti-AFB1 was added to the resulting sample. The extracted sample/antibody mix was passed over a waveguide surface patterned with immobilized AFB1. The resulting fluorescence signal decreased as the concentration of AFB1 in the sample increased. The limit of detection for AFB1 in buffer, 0.3 ng/mL, was found to increase to between 1.5 and 5.1 ng/g and 0.6 and 1.4 ng/g when measured in various corn and nut products, respectively.

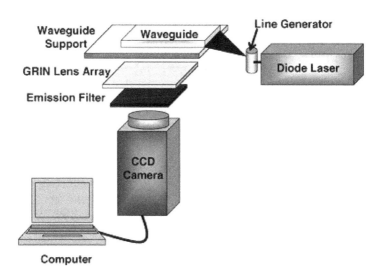

Fig. 8. Schematic of the NRL array biosensor. Reprinted from ref. 22 with permission by the Elsevier.

Ligler et al.[4] has clearly demonstrated the versatility of the biosensor arrays for the detection of both large and small food contaminants either individually or simultaneously. The bacterial pathogen C. jejuni was measured using a sandwich immunoassay both in buffer and additional complex food matrices with LODs ranging from 500 to 3780 CFU/ml. The mycotoxins OTA, DON, AFB1 and FB were detected simultaneously on a signal substrate using a competitive-based immunoassay format taking only 15min. The combination of sandwich and competitive immunoassay formats on a single substrate was demonstrated, allowing the simultaneous detection of both large (C. jejuni) and small (AFB1) food pathogens with LODs in buffer of 500 CFU/ml and 0.3 ng/mL, respectively.

Deoxynivalenol (DON), a mycotoxin produced by several Fusaruim species, is a worldwide contaminant of foods and feeds. Because of the potential dangers due to accidental or intentional contamination of foods with DON, there is a need to develop a rapid and highly sensitive method for easy identification and quantification of DON. In Taitt's study[23], they have developed and utilized a competitive immunoassay technique to detect DON in various food matrixes and indoor air samples using a biosensor array. A DON biotin conjugate, immobilized on a NeutrAvidin-coated optical waveguide, competed with the DON in the sample for binding to fluorescently labeled DON monoclonal antibodies. To demonstrate a simple procedure amenable for on-site use, DON-spiked cornmeal, cornflakes, wheat, barley, and oats were extracted with methanol water (3:1) and assayed without cleanup or pre-concentration. The limits of detection ranged from 0.2 ng/mL in buffer to 50 ng/g in oats. The detection limit of DON spiked into an aqueous effluent from an air sampler was 4 ng/mL.

Parro et al.[24] have developed antibodies and a multi-array competitive immunoassay (MACIA) for the detection of a wide range of molecular size compounds, from single aromatic ring derivatives or polycyclic aromatic hydrocarbons (PAHs), through small peptides, proteins or whole cells (spores). Multiple biosensor arrays containing target molecules are used simultaneously to run several competitive immunoassays. The sensitivity of the MACIA for small organic compounds like naphthalene, 4-phenylphenol or 4-terbutylphenol is in the range of 100–500 ppb, for others like the insecticide terbutryn it is at the ppt level, while for small peptides, as well as for more complex molecules like the protein thioredoxin, the sensitivity is approximately 1-2 ppb, or 104-105 spores of Bacillus subtilis per milliliter. For organic compounds, a water–methanol solution was used in order to achieve a better dissolution of the organics without compromising the antibody–antigen interaction. The above-mentioned compounds were detected by MACIA in water-(10%) methanol extracts from spiked pyrite and hematite-containing rock powder samples, as well as from a spiked-sand sample subjected to organic extraction with dichloromethane-methanol (1/1).

Sandwich immunoassays have also been conducted by running the samples through a multiple channel (one per flow channel) over the array of capture antibodies for a period of time sufficient for the antibody to bind any target agent in the sample. In most of the assays described by Ligler et al.[25], they use 8–10 min for binding in order to balance assay sensitivity with keeping the total assay time short (under 15 min). The sensitivity is clearly greater if longer periods of time are used, particularly if the samples are viscous and the rate of diffusion of the target within the sample is slow. In this array, biotinylated capture antibodies were exposed to the avidin-coated waveguide using flow channels molded out of polydimethylsiloxane as described previously. TNT (trinitrotoluene) antibody spots bound a Cy5-labeled TNB (trinitrobenzene) tracer molecule in the tracer cocktail to provide a positive control. Assays were conducted using staphylococcal enterotoxin B (SEB), ricin toxin, cholera toxin, mouse IgG and Bacillus globigii (B. globigii is an anthrax spore simulant) as targets. The Cy5-TNB bound to the appropriate antibodies in all lanes. Samples containing SEB, cholera toxin, mouse IgG and B. globigii bound cleanly to the appropriate spots without showing any cross-reactivity against other capture antibodies.

Contamination of food by mycotoxins occurs in minute quantities, and therefore, there is a need for a highly sensitive and selective device that can detect and quantify these organic toxins. Taitt et al.[26] reported the development of a rapid and highly sensitive array biosensor for the detection and quantitation of ochratoxin A (OTA). The array biosensor utilizes a competitive immunoassay format. Immobilized OTA derivatives compete with toxin in solution for binding to fluorescent anti-OTA antibody spiked into the sample. This competition is quantified by measuring the formation of the fluorescent immuno-complex on the waveguide surface. The fluorescent signal is inversely proportional to the concentration of OTA in the sample. Analyses for OTA in buffer and a variety of food and beverage samples were performed. Samples were extracted with methanol, without any sample cleanup or pre-concentration step prior to analysis. The limit of detection for OTA in several cereals ranged from 3.8 to 100 ng/g, while in coffee and wine, detection limits were 7 and 38 ng/g, respectively.

Golden et al.[27] have developed a "do-it-yourself" biosensor array for the simultaneous detection of multiple targets in multiple samples within 15-30 min. The biosensor is based on a planar waveguide, a modiWed microscope slide, with a pattern of small (mm²) sensing regions. The waveguide is illuminated by launching the emission of a 635 nm diode laser into the proximal end of the slide via a line generator. The evanescent Weld excites Xuorophores bound in the sensing region and the emitted Xuorescence is measured using a Peltier-cooled CCD camera. Assays can be performed on the waveguide in multichannel Xow chambers and then interrogated using the detection system described in their paper. This biosensor can detect many different targets, including proteins, toxins, cells, virus, and explosives with detection limit rivaling those of the ELISA detection system.

Kramer et al.[28] presents a new, versatile, portable miniaturized flow-injection biosensor array which is designed for field analysis. The temperature-controlled field prototype can run for 6 h without external power supply. The bio-recognition element is an analyte-specific antibody immobilized on a gold surface of pyramidal structures inside an exchangeable single-use chip, which hosts also the enzyme-tracer and the sample reservoirs. The competition between the enzyme-tracer and the analyte for the antigen-binding sites of the antibodies yields in the final step a chemiluminescence signal that is inversely proportional to the concentration of analyte in the given range of detection. A proof of principle is shown for nitroaromatics and pesticides. The detection limits reached with the field prototype in the laboratory was below 0.1 g/L for 2,4,6-trinitrotoluene (TNT), and about 0.2 g/L for diuron and atrazine, respectively. Important aspects in this development were the design of the competition between analyte and enzyme-tracer, the unspecific signal due to unspecific binding and/or luminescence background signal, and the flow pattern inside the chip.

The multianalyte array biosensor (MAAB) is a rapid analysis instrument capable of detecting multiple analytes simultaneously. Rapid (15 min), single-analyte sandwich immunoassays were developed for the detection of Salmonella enterica serovar Typhimurium by Taitt et al.[29], with a detection limit of 8×10^4 CFU/mL; the limit of detection was improved 10 fold by lengthening the assay protocol to 1 h. S. enterica serovar Typhimurium was also detected in the following spiked foodstuffs, with minimal sample preparation: sausage, cantaloupe, whole liquid egg, alfalfa sprouts, and chicken carcass rinse. Cross-reactivity tests were performed with *Escherichia coli* and *Campylobacter jejuni*. To determine whether the MAAB has potential as a screening tool for the diagnosis of asymptomatic Salmonella infection of poultry, chicken excretal samples from a private, noncommercial farm and from university poultry facilities were tested. While the private farm excreta gave rise to signals significantly above the buffer blanks, none of the university samples tested positive for S. enterica serovar Typhimurium without spiking; dose-response curves of spiked excretal samples from university-raised poultry gave limits of detection of 8×10^3 CFU/g.

Campylobacter and Shigella bacteria are common causes of food- and water-borne illness worldwide. There is a current need in food, medical, environmental, and military markets for a rapid and user-friendly method of detecting such pathogens. The array biosensor developed at the NRL encompasses these qualities. Ligler et al.[30] reported on a sandwich immunoassay-based biosensor array that was developed for the detection of Campylobacter and Shigella species in both buffer and a variety of food and beverage samples. The limit of detection for Shigella dysenteriae in buffer and chicken carcass wash was 4.9×10^4 CFU/mL, whereas Campylobacter jejuni could be measured at concentrations as low as 9.7×10^2 CFU/mL. The limits of detection and dynamic range were found to vary depending on the sample matrix, but could be improved by running the sample over the waveguide surface for longer periods of time. Samples were analyzed with no pre-concentration or enrichment steps and little-to-no sample pretreatment prior to analysis, and the total analysis run time was 25 min.

Biosensor array that is capable of detecting multiple targets rapidly and simultaneously on the surface of a single waveguide has also been studied. Ligler et al.[31] developed a sandwich and competitive fluoroimmunoassays to detect high and low molecular weight toxins, respectively, in complex samples. Antibodies were first immobilized in specific locations on the waveguide and the resultant patterned array was used to interrogate up to 12 different samples for the presence of multiple different analytes. Upon binding of a fluorescent analyte or fluorescent immunocomplex, the pattern of fluorescent spots was detected using a CCD camera. Automated image analysis was used to determine a mean fluorescence value for each assay spot and to subtract the local background signal. The location of the spot and its mean fluorescence value were used to determine the toxin identity and concentration. Toxins were measured in clinical fluids, environmental samples and foods, with minimal sample preparation. Results were reported for rapid analyses of staphylococcal enterotoxin B, ricin, cholera toxin, botulinum toxoids, trinitrotoluene, and the mycotoxin fumonisin. Toxins were detected at levels as low as 0.5 ng/mL.

4. Aptamer based biosensor array

Aptamers are single-stranded (ss) DNA or RNA molecules, typically <100 monomer units, which have the ability to bind to other molecules with high affinity and specificity. They are selected from random oligonucleotide pools by a process called Systematic Evolution of Ligands by Exponential enrichment (SELEX). Conceptually, the SELEX process is based on the ability of these small oligonucleotides to fold into unique three-dimensional (3-D) structures which can interact with a specific target with high specificity and affinity through such interactions as van der Waals surface contacts, hydrogen bonding and base stacking interactions. Aptamers can offer a strong and reliable role as biological recognition elements in most analytical applications. The specificity and high affinity of aptamers to a wide variety of targets, coupled with the ease of design and molecular engineering, as outlined earlier, make aptamers highly suitable for development as molecular biosensors. By moving in this direction, investigators have made appreciable efforts in recent years to utilize these unique features of aptamers to devise appropriate strategies with which to effectively and efficiently apply aptamers for biological agent recognition, identification, characterization and quantification[32].

Analogous to immunoassays based on the antigen-antibody interaction, aptamer-based biosensors can adopt different assay configurations to transduce bio-recognition events. Since aptamers have been selected to bind very different targets, ranging from small molecules to macromolecules, such as proteins, various assay configurations have been designed and reported. Nevertheless, the majority of these designs fall into two categories of configuration (Figure 9): single-site binding and dual-site binding[33]. The use of aptamers as new biological receptors in biosensor arrays can accelerate the development of biosensors of practical relevance. Because of their exceptionally high stability, selectivity and sensitivity, aptamer-based biosensor arrays have the potential to overcome the lacking functional and storage stability of most biosensors[34]. For example, an electrochemical impedance spectroscopy method of detection for aptamer-based electrochemical biosensor array (Figure 10) is reported in which the binding of aptamers immobilized on gold electrodes leads to impedance changes associated with target protein binding events by Xu et al.[35]. Human IgE was used as a model target protein and incubated with the aptamer-based array consisting of single-stranded DNA containing a hairpin loop. To increase the binding efficiency for proteins, a hybrid modified layer containing aptamers and cysteamine was fabricated on the photolithographic gold surface through molecular self-assembly. Compared to immunosensing methods using anti-human IgE antibody as the recognition element, impedance spectroscopy detection could provide higher sensitivity and better selectivity for aptamer-modified electrodes. The results of this method show good correlation for human IgE in the range of 2.5-100 nmol/L. A detection limit of 0.1 nmol/L was obtained, and an average of the relative standard deviation was 10%. The method describes the first label-free detection for arrayed electrodes utilizing electrochemical impedance spectroscopy.

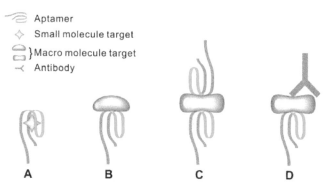

Fig. 9. Aptamer-based assay formats. (A) Small-molecule target buried within the binding pockets of aptamer structures; (B) single-site format;(C) dual-site (sandwich) binding format with two aptamers; and, (D) "sandwich" binding format with an aptamer. Reprinted from ref. 33 with permission by the Elsevier.

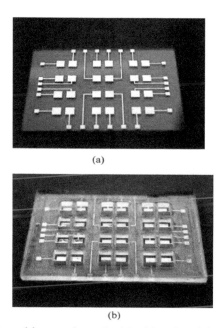

Fig. 10. Construction of the gold array electrode chip: (a) a photolithographic gold film array electrode; (b) a home-constructed PDMS frame containing 24-microwells for the immobilization. Reprinted from ref. 35 with permission by the Royal Society of Chemistry.

Quandt et al.[36] designed a Love-wave biosensor array by coupling aptamers to the surface of a Love-wave sensor chip. The sensor chip consists of five single sensor elements and allows label-free, real-time, and quantitative measurements of protein and nucleic acid binding events in concentration-dependent fashion. The biosensor was calibrated for human-thrombin and HIV-1 Rev peptide by binding fluorescently labeled molecules and correlating the mass of the bound molecules to fluorescence intensity. Detection limits of approximately 75 pg/cm² were obtained, and analyte recognition was specific. The sensor can easily be regenerated by simple washing steps. They further demonstrated the versatile applicability of the sensor by immobilizing single-stranded DNA (ssDNA) for the detection of the corresponding counter-strand.

The large quantity of aptamers which have been selected to bind complex molecules of low molecular weight leads to the possible use of these aptamers not only in diagnostic assays, but also in a wider range of applications, such as environmental analytical chemistry[37]. Selection of DNA ligands to the chloroaromatics, 4-chloroaniline (4-CA), 2,4,6-trichloroaniline (TCA) and pentachlorophenol (PCP), was performed by a novel method utilizing magnetic beads (MBs) having a linker arm for immobilization[38]. Moreover, Labuda et al.[39] reported for the first time the selection of RNA aptamers for the recognition of hydrophobic aromatic carcinogens. In particular, RNA aptamers with a K_d in the low micro-molar range have been selected for aromatic amines residues using as a model methylendianiline, which is a common industrial chemical employed to manufacture plastics, glues and foams.

A toxin-related work based on aptamers arrays have been published by Ellington et al.[40] . The authors reported the adaptation of a chip-based micro-sphere array (the "electronic taste chip") to aptamer receptors. Their detection system is illustrated in Figure 11. Unlike most protein-based arrays, the aptamer chips could be stripped and reused multiple times. The aptamer chips proved to be useful for screening aptamers from in vitro selection experiments and for sensitively quantitating the bio-threat agent ricin. The system composed of a flow cell connected to a fast performance liquid chromatography pump and a fluorescence microscope for observation. The flow cells contained silicon chips with multiple wells in which beads modified with the sensor elements were deposited. Commercially available streptavidin agarose beads were modified with biotinylated aptamers; RNA anti-ricin aptamers were used to demonstrate the possibility of quantifying the labeled protein. A sandwich assay format was also optimized using anti-ricin antibodies, to directly detect the unlabelled protein. In the first type of assay, the aptamer was bio-tinylated, immobilized and put in contact with the solution containing fluorescently labeled ricin, once introduced into the chip wells. The fluorescence intensities of the captures proteins were used to construct a calibration plot for ricin and a detection limit of 8 mg/ml was obtained. In the sandwich assay, the anti-ricin aptamer acted as a capture reagent and unlabelled ricin bound to the aptamer could interact with fluorophore-labeled fabricated an aptamer-based biosensor array for protein detection.

Environmental allergenic disease is a major cause of illness and disability, and there is broad consensus that the prevalence of type I allergy is increasing worldwide. Recent advances in biotechnology have yielded potentially useful functional binding aptamers that can enable low cost, high affinity allergen measurement. Aptamers are selected in vitro from combinatorial oligonucleotide libraries and therefore have several advantages over the traditionally used antibodies for detection of allergens. Aptamer-based methods could be used for measuring environmental allergens. Integrating the resulting aptamer-based allergen measurements to enhance quantization in an ongoing and complementary environmental childhood asthma epidemiological study forms the basis for the third and final aim. Successful use of aptamers for measuring environmental allergens should lead to a more cost effective, flexible, and health relevant method and thereby provides the potential for a more fundamental understanding of the role of environmental allergens in respiratory health.

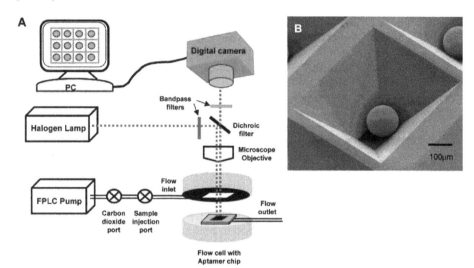

Fig. 11. Detection systems. (A) The electronic tongue setup contains a fluid delivery system, fluorescence microscope, digital camera, flow cell in which the aptamer chip will be loaded, and computer for data analysis. (B) Close-up look at a bead in a rectangular-shaped micro-machined well of the aptamer chip. Reprinted from ref. 40 with permission by the Royal Society of Chemistry.

5. Enzyme based biosensor array

Enzyme-based technology relies upon the natural specificity of given enzymatic protein to react biochemically with a target substrate or substrates. Like ion channels, there are many enzymes that participate in cellular signaling and, in some cases, are targeted by compounds associated with environmental toxicity. In general, enzyme-based biosensors employ semi-permeable membranes through which target analytes diffuse toward a solid-phase immobilized enzyme compartment. Ion selective, amperometric, or pH electrodes measure reaction components such as hydrogen peroxide (from oxidation of glucose by glucose oxidase) or ammonium ions (from urease metabolism of urea)[41]. Enzymes were historically the first molecular recognition elements included in biosensors and continue to be the basis for a significant number of publications reported for biosensors in general as well as biosensors for environmental applications. There are several advantages for enzyme biosensors. These include a stable source of material (primarily through bio-renewable sources), the ability to modify the catalytic properties or substrate specificity by means of genetic engineering, and catalytic amplification of the biosensor response by modulation of the enzyme activity with respect to the target analyte[17].

Recent progress with respect to enzyme biosensors for environmental applications has been reported in several areas[42]. These areas include the following: genetic modification of enzymes to increase assay sensitivity, stability and shelf life; improved electrochemical interfaces and mediators for more efficient operation; and introduction of sampling schemes consistent with potential environmental applications. More recently, enzyme-based biosensor arrays also have been used in the application of environmental monitoring. For example, Kukla et al.[43] developed a multi-enzyme electrochemical biosensor array. Their sensor array is based on capacitance pH-sensitive electrolyte–insulator–semiconductor (EIS) sensors with silicon nitride ion-sensitive layers and different forms of cholinesterase, urease and glucose oxidase as sensitive elements. With this sensor array, the authors used a multi-enzyme analysis to recognize the heavy metal ions in solutions containing a mixture of different metal ions, as well as for determination of the metal ion content in the analyzed samples. The content of toxic elements was determined by estimation of the residual activity of enzymatic membranes after the injection of analyzed samples. The conditions for enzyme sensors operation, such as buffer capacity, substrate concentration, time of incubation and time of response signal measurement, were optimized to reach the maximal sensitivity of multi-sensor for analysis of heavy metal ions in the investigated solutions. The results show that multi-enzyme analysis followed by mathematical processing is an efficient approach to develop biosensor arrays for toxic substrates detection.

Organophosphate pesticides (OPs) used to be widely used in agriculture due to their high efficiency as insecticides. OPs have been shown to result in high levels of acute neuron-toxicity and carcinogenicity, with the majority being hazardous to both human health and to the wider environment. A rapid, reliable, economical and portable analytical system will be of great benefit in the detection and prevention of OPs contamination. A biosensor array based on six acetylcholinesterase enzymes coupled with a novel automated instrument incorporating a neural network program has been reported by Hart et al.[44]. The biosensor array and the instrument is illustrated in Figure 12. Electrochemical analysis was carried out using chronoamperometry and the measurement was taken 10 s after applying a potential of 0 V vs. Ag/AgCl. The total analysis time for the complete assay was less than 6 min. The array was used to produce calibration data with six organophosphate pesticides (OPs) in the concentration range of 10^{-5} mol/L to 10^{-9} mol/L to train a neural network. The output of the neural network was subsequently evaluated using different sample matrices. There was no detrimental matrix effect observed from water, phosphate buffer, food or vegetable extracts. Furthermore, the sensor system was not detrimentally affected by the contents of water samples taken from each stage of the water treatment process. Their biosensor array system successfully identified and quantified all samples where an OP was present in water, food and vegetable extracts containing different OPs. There were no false positives or false negatives observed during the evaluation of the analytical system. Their biosensor arrays and automated instrument were evaluated in situ in field experiments where the instrument was successfully applied to the analysis of a range of environmental samples.

Recently, many studies have focused on the development of biochemical sensors, which are well suited for the rapid, simple and selective analysis of pesticides. Specially, they combine the selectivity of the enzymatic reactions with operational simplicity and simple detection schemes. Valle et al.[45] developed an electronic tongue, employing an array of inhibition biosensors and Artificial Neural Networks (ANNs). The array of biosensors was made up of three amperometric pesticide biosensors that used different acetylcholinesterase (AChE) enzymes: a wild type from electric eel (EE) and two different genetically modified enzymes

(B1 and B394). In order to model the response to dichlorvos and carbofuran mixtures, a total amount of 22 solutions were prepared, with random concentrations. Chronoamperometric responses of the biosensor array were used in order to obtain the inhibition bioelectronic tongue. Mean values of concentration of pesticides evaluated were 0.79 nmol/L for dichlorvos and 4.1 nmol/L for carbofuran. Good prediction ability was obtained with correlation coefficients better than 0.918 when the obtained values were compared with those expected for a set of 6 external test samples not used for training.

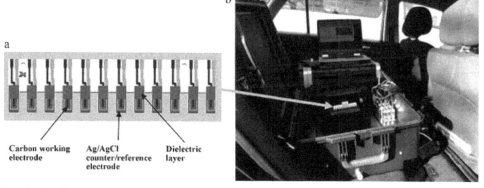

Fig. 12. (a) electrode array comprising 12 screen-printed carbon electrodes and an Ag/AgCl counter/reference electrode printed on an alumina substrate; (b) array in the prototype biosensor system operating in the field powered from a car battery via the lighter socket. Reprinted from ref. 44 with permission by the Elsevier.

Another approach is by using Organophosphorus hydrolase (OPH). OPH is a 72 kDa homodimeric, metalloenzyme, containing two zinc ions in the active site involved in catalytic and/or structural functions. OPH catalyzes the hydrolysis of Organophosphates (OPs) resulting in its detoxification. Some of the biosensors that were developed exploiting OPH as the bio-recognition element on different detection platforms have been reported. Though highly sensitive and selective towards different OPs, their inability to provide simultaneous measurements of different analytes was a major shortcoming. Simonian et al.[46] developed a biosensor array (Figure 13) with the potential for direct detection of organophosphates using OPH, conjugated with a pH-sensitive fluorophore, carboxynaphthofluorescein (CNF). The presence of reference spots allows the discrimination of the enzymatic and non-enzymatic based pH changes; bovine serum albumin (BSA) was used as a non-enzymatic scaffold protein for CNF attachment at the reference spots. An array biosensor unit developed at the Naval Research Laboratories (NRL) was adopted as the detection platform and appropriately modified for enzyme-based measurements. A planar multi-mode waveguide was covered with an optically transparent TiO_2 layer to increase the surface area available for immobilization. The biosensor enabled the detection of 2.5 µmol/L paraoxon, and 10 µmol/L parathion respectively. Very short response time of 30 s can be achieved with a total analysis time of less than 2 min. When operated at room temperature and stored at 4 °C, the waveguide retained reasonable activity for greater than 45 days.

An array-based optical biosensor for the simultaneous analysis of multiple samples in the presence of unrelated multi-analytes was fabricated by Doong et al.[47]. The authors used

Urease and acetylcholinesterase (AChE) as model enzymes, which were co-entrapped with the sensing probe, FITC-dextran, in the sol-gel matrix to measure pH, urea, acetylcholine (ACh) and heavy metals (enzyme inhibitors). Environmental and biological samples spiked with metal ions were also used to evaluate the application of the array biosensor to real samples. The biosensor exhibited high specificity in identifying multiple analytes. No obvious cross-interference was observed when a 50-spot array biosensor was used for simultaneous analysis of multiple samples in the presence of multiple analytes. The sensing system can determine pH over a dynamic range from 4 to 8.5. The limits of detection of 2.5-50 µmol/L with a dynamic range of 2-3 orders of magnitude for urea and ACh measurements were obtained. Moreover, the urease-encapsulated array biosensor was used to detect heavy metals. The analytical ranges of Cd(II), Cu(II), and Hg(II) were between 10 nmol/L and 100 mmol/L. When real samples were spiked with heavy metals, the array biosensor also exhibited potential effectiveness in screening enzyme inhibitors.`

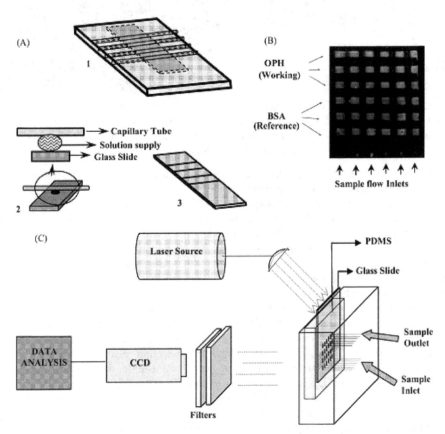

Fig. 13. (A) Schematic of modified process for incubation using thin glass tubes. (B) Schematic of the glass slide with immobilized proteins and fluorophores. (C) Schematic of the array biosensor. Reprinted from ref. 46 with permission by the Elsevier.

Solna et al.[48] use screen-printed four-electrode system as the amperometric transducer for determination of phenols and pesticides using immobilized tyrosinase, peroxidase, acetylcholinesterase and butyrylcholinesterase. Acetylthiocholine chloride was chosen as substrate for cholinesterases to measure inhibition by pesticides, hydrogen peroxide served as co-substrate for peroxidase to measure phenols. In their work, the compatibility of hydrolases and oxidoreductases working in the same array was studied. The detection of p-cresol, catechol and phenol as well as of pesticides including carbaryl, heptenophos and fenitrothion was carried out in flow-through and steady state arrangements. It was demonstrated that electrodes modified with hydrolases and oxidoreductases can function in the same array. The limit of detection for catechol using tyrosinase was equal to 0.35 and 1.7 µmol/L in the flow and steady systems. Lower limits of detection for pesticides were achieved in the steady state system: carbaryl 26 nmol/L, heptenophos 14 nmol/L and fenitrothion 0.58 nmol/L. Similar multi-enzyme-based electrochemical biosensor arrays for the determination of pesticides[49-52] and phenols[53] have been reported by other workers.

6. Microorganism-based biosensor array

A microbial biosensor is an analytical device which integrates microorganism(s) with a physical transducer to generate a measurable signal proportional to the concentration of analytes. In recent years, a large number of microbial biosensors have been developed for environmental, food, and biomedical applications[54].

Enzymes are the most widely used biological sensing element in the fabrication of biosensors. Although purified enzymes have very high specificity for their substrates or inhibitors, their application in biosensors construction may be limited by the tedious, time-consuming and costly enzyme purification, requirement of multiple enzymes to generate the measurable product or need of cofactor/coenzyme. Microorganisms provide an ideal alternative to these bottle-necks. The many enzymes and co-factors that co-exist in the cells give the cells the ability to consume and hence detect large number of chemicals; however, this can compromise the selectivity. They can be easily manipulated and adapted to consume and degrade new substrate under certain cultivating condition. Additionally, the progress in molecular biology/recombinant DNA technologies has opened endless possibilities of tailoring the microorganisms to improve the activity of an existing enzyme or express foreign enzyme/protein in host cell. All of these make microbes excellent biosensing elements[55].

Microorganism-based biosensor arrays classically used for environmental biosensing are mainly bacteria and yeasts, and to a lesser extent algae. Various strains have been exploited, from commercial and well-characterized cells harboring a broad range of substrates to genetically engineered organisms specially constructed to detect specific molecules or groups of molecules, passing through environmental cells isolated from polluted sites offering greater robustness and more specific enzymatic properties[56].

Rapid identification of *Escherichia coli* strains is an important diagnostic goal in applied medicine as well as the environmental and food sciences. Mikkelsen et al.[57] reported an electrochemical, screen-printed biosensor array, where selective recognition is accomplished using lectins that recognize and bind to cell-surface lipopolysaccharides and coulometric transduction exploits non-native external oxidants to monitor respiratory cycle activity in lectin-bound cells. Ten different lectins were separately immobilized onto porous membranes that feature activated surfaces. Modified membranes were exposed to untreated *E. coli* cultures for 30 min, rinsed, and layered over the individual screen-printed carbon electrodes of the sensor array. The membranes were incubated 5 min in a reagent solution that contained the oxidants menadione and ferricyanide as well as the respiratory substrates succinate and formate. Electrochemical oxidation of ferrocyanide for 2 min provided chronocoulometric data related to the quantities of bound cells. These screen-printed sensor arrays were used in conjunction with factor analysis for the rapid identification of four *E. coli* subspecies (*E. coli* B, *E. coli* Neotype, *E. coli* JM105 and *E. coli* HB101). Systematic examination of lectin-binding patterns showed that these four *E. coli* subspecies are readily distinguished using only five essential lectins.

The last decade has witnessed a significant increase in interest in whole-cell biosensors for diverse applications, as well as a rapid and continuous expansion of array technologies. The combination of these two disciplines has yielded the notion of whole-cell array biosensors. Belkin et al.[58] presented a potential manifestation of this idea by describing the printing of a whole-cell bacterial bioreporters array (Figure 14). Exploiting natural bacterial tendency to adhere to positively charged abiotic surfaces, they describe immobilization and patterning of bacterial "spots" in the nanoliter volume range by a non-contact robotic printer. They show that the printed Escherichia coli-based sensor bacteria are immobilized on the surface, and retain their viability and biosensing activity for at least 2 months when kept at 4°C. Immobilization efficiency was improved by manipulating the bacterial genetics, the growth and the printing media and by a chemical modification of the inanimate surface. The result suggests that the methodology presented by them may be applicable to the manufacturing of whole-cell sensor arrays for diverse high throughput applications. In the course of the study, they have also described a novel specific reporter for the detection of respiratory inhibitors. Sodium azide, a chemical with a constantly increasing world distribution, served as the model toxicant. The sensor's response was rapid (20 minutes after exposure) and dose-dependent, and could be maintained for at least 2 months at 4 °C.

Li et al.[59] developed a double interdigitated array microelectrodes (IAM)-based flow cell for an impedance biosensor to detect viable *Escherichia coli* O157:H7 cells after enrichment in a growth medium. Their study was aimed at the design of a simple flow cell with embedded IAM which does not require complex microfabrication techniques and can be used repeatedly with a simple assembly/disassembly step. The flow cell was also unique in having two IAM chips on both top and bottom surfaces of the flow cell, which enhances the sensitivity of the impedance measurement. *E. coli* O157:H7 cells were grown in a low conductivity yeast–peptone–lactose–TMAO (YPLT) medium outside the flow cell. After bacterial growth, impedance was measured inside the flow cell. Equivalent circuit analysis indicated that the impedance change caused by bacterial growth was due to double layer capacitance and bulk medium resistance. Both parameters were a function of ionic concentration in the medium, which increased during bacterial growth due to the conversion of weakly charged substances present in the medium into highly charged ions. The impedance biosensor successfully detected *E. coli* O157:H7 in a range from 8.0 to 8.2×10^8 CFU/mL after an enrichment growth of 14.7 and 0.8 h, respectively. A logarithmic linear relationship between detection time (T_D) in h and initial cell concentration (N_0) in CFU/mL was $T_D = -1.73 \log N_0 + 14.62$, with $R^2 = 0.93$. Double IAM-based flow cell was more sensitive than single IAM-based flow cell in the detection of *E. coli* O157:H7 with 37–61% more impedance change for the frequency range from 10 Hz to 1 MHz. The double IAM-based flow cell could be used to design a simple impedance biosensor for the sensitive detection of bacterial growth and their metabolites.

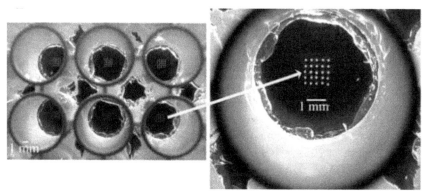

Fig. 14. Twenty five spots, 1 nl each, of strain SM118 in ectoine, printed onto the wells of 96-well plate with an APTES coated glass bottom. Reprinted from ref. 58 with permission by the Royal Society of Chemistry.

Worldwide herbicide discharge into the aquatic environment is also a growing concern. Adverse effects induced by herbicide contamination are impacting a great variety of organisms and ecosystems, ranging from the primary producers to animals and humans. Biosensors for the rapid detection of herbicides in the environment have also been explored. A multiple-strain algal biosensor was constructed for the detection of herbicides inhibiting photosynthesis by Podola et al.[60]. Nine different microalgal strains were immobilized on an array biochip using permeable membranes. The biosensor allowed on-line measurements of aqueous solutions passing through a flow cell using chlorophyll fluorescence as the biosensor response signal. The herbicides atrazine, simazine, diuron, isoproturon and paraquat were detectable within minutes at minimal LOEC (Lowest Observed Effect Concentration) ranging from 0.5 to 100 μg/L, depending on the herbicide and algal strain. The most sensitive strains in terms of EC50 values were Tetraselmis cordiformis and Scherffelia dubia. Less sensitive species were Chlorella vulgaris, Chlamydomonas sp. and Pseudokirchneriella subcapitata, but for most of the strains no general sensitivity or resistance was found. The different responses of algal strains to the five herbicides constituted a complex response pattern (RP), which was analyzed for herbicide specificity within the linear dose-response relationship.

Recombinant bioluminescent bacterial strains are increasingly receiving attention as environmental biosensors due to their advantages, such as high sensitivity and selectivity, low costs, ease of use and short measurement times. Gu et al.[61] use a cell-based array technology that uses recombinant bioluminescent bacteria to detect and classify environmental toxicity followed by developing two biosensor arrays, i.e., a chip and a plate array. Twenty recombinant bioluminescent bacteria, having different promoters fused with the bacterial lux genes, were immobilized within LB-agar. About 2 μl of the cell-agar mixture was deposited into the wells of either a cell chip or a 384-well plate. The bioluminescence (BL) from the cell arrays was measured with the use of highly sensitive cooled CCD camera that measured the bioluminescent signal from the immobilized cells and then quantified the pixel density using image analysis software. The responses from the cell arrays were characterized using three chemicals that cause either superoxide damage (paraquat), DNA damage (mitomycin C) or protein/membrane damage (salicylic acid). The responses were found to be dependent upon the promoter fused upstream of the lux operon within each strain. Therefore, a sample's toxicity can be analyzed and classified through the changes in the BL expression from each well. Moreover, a time of only 2 h was needed for analysis, making either of these arrays a fast, portable and economical high-throughput biosensor system for detecting environmental toxicities.

Because of their ability to perform functional sensing, living cell-based biosensors are drawing increased attention. The work reported by Walt et al.[62] demonstrates the ability to fabricate an optical imaging fiber-based living bacterial cell array for genotoxin detection. A biosensor composed of a high-density living bacterial cell array was fabricated by inserting bacteria into a micro-well array formed on one end of an imaging fiber bundle. The size of each micro-well allows only one cell to occupy each well. In this biosensor, E. coli cells carrying a recA::gfp fusion were used as sensing components for genotoxin detection. Each fiber in the array has its own light pathway, enabling thousands of individual cell responses to be monitored simultaneously with both spatial and temporal resolution. The biosensor was capable of performing cell-based functional sensing of a genotoxin with high sensitivity and short incubation times (1 ng/mL mitomycin C after 90 min). The biosensors demonstrated an active sensing lifetime of more than 6 h and a shelf lifetime of two weeks. Their group reported another live cell biosensor array[63], which was fabricated by immobilizing bacterial cells on the face of an optical imaging fiber containing a high density array of micro-wells. Each microwell accommodates a single bacterium that was genetically engineered to respond to a specific analyte. A genetically modified Escherichia coli strain, containing the lacZ reporter gene fused to the heavy metal-responsive gene promoter zntA, was used to fabricate a mercury biosensor. A plasmid carrying the gene coding for the enhanced cyan fluorescent protein (ECFP) was also introduced into this sensing strain to identify the cell locations in the array. Single cell lacZ expression was measured when the array was exposed to mercury and a response to 100 nmol/L Hg^{2+} could be detected after a 1-h incubation time. The optical imaging fiber-based single bacterial cell array is a flexible and sensitive biosensor platform that can be used to monitor the expression of different reporter genes and accommodate a variety of sensing strains.

7. Conclusion and future direction

In recent years, there have been dramatic advances in a new analytical format, the biosensor array, a tool that has revolutionized our ability to characterize and quantify biologically and envitonmetally relevant molecules. The biosensor arrays address the need for rapid, sensitive, and specific screening for multiple pollutants at the site of sample collection. The biosensor arrays have several very significant advantages for such applications: (1) The number of analyte which can be detected simultaneously can be expanded as need dictates and specific analyte become available. (2) The biosensor arrays and tracer reagents are reusable if no target agent binds to the array surface. This feature significantly decreases the cost and operational burden for the user and simplifies automation for extended monitoring applications. (3) The biosensor array is simple to use. It is easily portable for first responder applications. The insertion of the sensor array, tracer reagents and samples is very simple with no requirement for alignment operations by the user. (4) The biosensor array is a low-cost system which can be made even more cost effective with mass production. (5) The biosensor array can be easily adapted for continuous monitoring operations by integration with a computer-controlled sampler to format automatic analytical system. Because of these advantages, more and more biosensor arrays are applied in varied areas including environmental monitoring. An overview of the applications for environment by using biosensor arrays, which are not mentioned in this review, are listed in Table 1.

Target	Biosensor array type	LOD	Reference
Herbicide Subclasses	Array of photosystem II mutants	3×10^{-9} mol/L	[64]
Metal ions	All-solid-state potentiometric biosensor array	10^{-6} mol/L	[65]
Microbial species	Electrochemical biosensor array	Not given	[66]
Escherichia coli	Quantum dot-based array	10 CFU/mL	[67]
Bio-hazardous agents	Planar waveguide biosensor array	5×10^{5} CFU/mL	[68]
aflatoxin B_1	NRL biosensor array	0.6 ng/g	[69]
Ochratoxin A	Antibody-based biosensor array	3.8 ng/g	[70]
Odour	Colorimetric biosensor array	Not given	[71]
Escherichia coli	Antimicrobial Peptides based biosensor array	10^{7} CFU/mL	[72]
Yersinia pestis F1	Antibody-based biosensor array	25 ng/mL	[73]
Bacillus globigii	Antibody-based biosensor array	10^{5} CFU/mL	[74]
Shigella dysenteriae	Antibody-based biosensor array	5×10^{4} CFU/mL	[75]

Table 1. Applications of biosensor arrays for environmental monitoring

Despite the high number of biosensor arrays under development and the amount of research literature on this area, few practical systems are currently enjoying market acceptance for environmental applications. The Naval Research Laboratory (NRL) biosensor arrays are the most successful type of biosensor arrays that have found commercial application not only in environmental monitoring but also in the monitoring of bio-molecular interaction events in general. Biosensor arrays still need more research and development in order to achieve the stability, sensitivity, specificity, and versatility that will attract confidence of potential users, especially for biotechnology and environmental applications.

8. References

[1] G. Hanrahan, D.G. Patil, J. Wang, J Environ Monitor, 6 (2004) 657-664.
[2] M. Badihi-Mossberg, V. Buchner, J. Rishpon, Electroanal, 19 (2007) 2015-2028.
[3] A. Cavalcanti, B. Shirinzadeh, M. Zhang, L. Kretly, Sensors, 8 (2008) 2932-2958.
[4] K.E. Sapsford, M.M. Ngundi, M.H. Moore, M.E. Lassman, L.C. Shriver-Lake, C.R. Taitt, F.S. Ligler, Sensor Actuat B-Chem, 113 (2006) 599-607.
[5] M. Seidel, R. Niessner, Analytical and Bioanalytical Chemistry, 391 (2008) 1521-1544.
[6] F.S. Ligler, K.E. Sapsford, J.P. Golden, L.C. Shriver-Lake, C.R. Taitt, M.A. Dyer, S. Barone, C.J. Myatt, Anal Sci, 23 (2007) 5-10.

[7] J. Zhai, H. Cui, R. Yang, Biotechnology Advances, 15 (1997) 43-58.

[8] A. Sassolas, B.D. Leca-Bouvier, L.J. Blum, Chemical Reviews, 108 (2007) 109-139.

[9] Y.-C. Cao, Z.-L. Huang, T.-C. Liu, H.-Q. Wang, X.-X. Zhu, Z. Wang, Y.-D. Zhao, M.-X. Liu, Q.-M. Luo, Anal Biochem, 351 (2006) 193-200.

[10] J. Wang, G. Liu, A. Merkoçi, J Am Chem Soc, 125 (2003) 3214-3215.

[11] H. Li, Z. Sun, W. Zhong, N. Hao, D. Xu, H.-Y. Chen, Analytical Chemistry, 82 (2010) 5477-5483.

[12] D.D. Zhang, Y.G. Peng, H.L. Qi, Q. Gao, C.X. Zhang, Biosens Bioelectron, 25 (2010) 1088-1094.

[13] E. Komarova, K. Reber, M. Aldissi, A. Bogomolova, Biosensors and Bioelectronics, 25 (2010) 1389-1394.

[14] H. Xia, F. Wang, Q. Huang, J.F. Huang, M. Chen, J. Wang, C.Y. Yao, Q.H. Chen, G.R. Cai, W.L. Fu, Sensors, 8 (2008) 6453-6470.

[15] R.A. Doong, H.M. Shih, S.H. Lee, Sensor Actuat B-Chem, 111 (2005) 323-330.

[16] N.J. Ronkainen, H.B. Halsall, W.R. Heineman, Chem Soc Rev, 39 (2010) 1747-1763.

[17] K.R. Rogers, Analytica Chimica Acta, 568 (2006) 222-231.

[18] W. Lian, D. Wu, D.V. Lim, S. Jin, Anal Biochem, 401 (2010) 271-279.

[19] K. Kloth, R. Niessner, M. Seidel, Biosensors and Bioelectronics, 24 (2009) 2106-2112.

[20] K. Kloth, M. Rye-Johnsen, A. Didier, R. Dietrich, E. Martlbauer, R. Niessner, M. Seidel, Analyst, 134 (2009) 1433-1439.

[21] M. Varshney, Y.B. Li, Biosens Bioelectron, 22 (2007) 2408-2414.

[22] K.E. Sapsford, C.R. Taitt, S. Fertig, M.H. Moore, M.E. Lassman, C.A. Maragos, L.C. Shriver-Lake, Biosens Bioelectron, 21 (2006) 2298-2305.

[23] M.M. Ngundi, S.A. Qadri, E.V. Wallace, M.H. Moore, M.E. Lassman, L.C. Shriver-Lake, F.S. Ligler, C.R. Taitt, Environmental Science & Technology, 40 (2006) 2352-2356.

[24] P. Fernández-Calvo, C. Näke, L.A. Rivas, M. García-Villadangos, J. Gómez-Elvira, V. Parro, Planet Space Sci, 54 (2006) 1612-1621.

[25] L.C. Shriver-Lake, F.S. Ligler, Ieee Sens J, 5 (2005) 751-756.

[26] M.M. Ngundi, L.C. Shriver-Lake, M.H. Moore, M.E. Lassman, F.S. Ligler, C.R. Taitt, Analytical Chemistry, 77 (2005) 148-154.

[27] J. Golden, L. Shriver-Lake, K. Sapsford, F. Ligler, Methods, 37 (2005) 65-72.

[28] I.M. Ciumasu, P.M. Krämer, C.M. Weber, G. Kolb, D. Tiemann, S. Windisch, I. Frese, A.A. Kettrup, Biosensors and Bioelectronics, 21 (2005) 354-364.

[29] C.R. Taitt, Y.S. Shubin, R. Angel, F.S. Ligler, Appl. Environ. Microbiol., 70 (2004) 152-158.

[30] K.E. Sapsford, A. Rasooly, C.R. Taitt, F.S. Ligler, Analytical Chemistry, 76 (2004) 433-440.

[31] F. Ligler, C. Taitt, L. Shriver-Lake, K. Sapsford, Y. Shubin, J. Golden, Analytical and Bioanalytical Chemistry, 377 (2003) 469-477.

[32] K. Sefah, J.A. Phillips, X. Xiong, L. Meng, D. Van Simaeys, H. Chen, J. Martin, W. Tan, Analyst, 134 (2009) 1765-1775.

[33] S. Song, L. Wang, J. Li, C. Fan, J. Zhao, TrAC Trends in Analytical Chemistry, 27 (2008) 108-117.

[34] B. Strehlitz, N. Nikolaus, R. Stoltenburg, Sensors, 8 (2008) 4296-4307.

[35] D. Xu, D. Xu, X. Yu, Z. Liu, W. He, Z. Ma, Analytical Chemistry, 77 (2005) 5107-5113.

[36] M.D. Schlensog, T.M.A. Gronewold, M. Tewes, M. Famulok, E. Quandt, Sensors and Actuators B: Chemical, 101 (2004) 308-315.

[37] S. Tombelli, M. Minunni, M. Mascini, Biomol Eng, 24 (2007) 191-200.

[38] J.G. Bruno, Biochemical and Biophysical Research Communications, 234 (1997) 117-120.

[39] U. Brockstedt, A. Uzarowska, A. Montpetit, W. Pfau, D. Labuda, Biochemical and Biophysical Research Communications, 313 (2004) 1004-1008.

[40] R. Kirby, E.J. Cho, B. Gehrke, T. Bayer, Y.S. Park, D.P. Neikirk, J.T. McDevitt, A.D. Ellington, Analytical Chemistry, 76 (2004) 4066-4075.

[41] J.J. Pancrazio, J.P. Whelan, D.A. Borkholder, W. Ma, D.A. Stenger, Ann Biomed Eng, 27 (1999) 697-711.

[42] S. Rodriguez-Mozaz, M. Lopez de Alda, D. Barceló, Analytical and Bioanalytical Chemistry, 386 (2006) 1025-1041.

[43] A.L. Kukla, N.I. Kanjuk, N.F. Starodub, Y.M. Shirshov, Sensors and Actuators B: Chemical, 57 (1999) 213-218.

[44] A. Crew, D. Lonsdale, N. Byrd, R. Pittson, J.P. Hart, Biosens Bioelectron, 26 (2011) 2847-2851.

[45] M. Cortina, M. del Valle, J.-L. Marty, Electroanal, 20 (2008) 54-60.

[46] M. Ramanathan, A.L. Simonian, Biosens Bioelectron, 22 (2007) 3001-3007.

[47] H.-c. Tsai, R.-a. Doong, Biosensors and Bioelectronics, 20 (2005) 1796-1804.

[48] R. Solná, S. Sapelnikova, P. Skládal, M. Winther-Nielsen, C. Carlsson, J. Emnéus, T. Ruzgas, Talanta, 65 (2005) 349-357.

[49] R. Solná, E. Dock, A. Christenson, M. Winther-Nielsen, C. Carlsson, J. Emnéus, T. Ruzgas, P. Skládal, Analytica Chimica Acta, 528 (2005) 9-19.

[50] S. Sapelnikova, E. Dock, R. Solná, P. Skládal, T. Ruzgas, J. Emnéus, Analytical & Bioanalytical Chemistry, 376 (2003) 1098-1103.

[51] J.J. Pancrazio, S.A. Gray, Y.S. Shubin, N. Kulagina, D.S. Cuttino, K.M. Shaffer, K. Eisemann, A. Curran, B. Zim, G.W. Gross, T.J. O'Shaughnessy, Biosensors and Bioelectronics, 18 (2003) 1339-1347.

[52] S.J. Young, J.P. Hart, A.A. Dowman, D.C. Cowell, Biosensors and Bioelectronics, 16 (2001) 887-894.

[53] S. Sapelnikova, E. Dock, T. Ruzgas, J. Emnéus, Talanta, 61 (2003) 473-483.

[54] L. Su, W. Jia, C. Hou, Y. Lei, Biosensors and Bioelectronics, 26 (2011) 1788-1799.

[55] Y. Lei, W. Chen, A. Mulchandani, Anal Chim Acta, 568 (2006) 200-210.

[56] F. Lagarde, N. Jaffrezic-Renault, Anal Bioanal Chem, 400 (2011) 947-964.

[57] P. Ertl, M. Wagner, E. Corton, S.R. Mikkelsen, Biosens Bioelectron, 18 (2003) 907-916.

[58] S. Melamed, L. Ceriotti, W. Weigel, F. Rossi, P. Colpo, S. Belkin, Lab Chip, 11 (2011) 139-146.

[59] M. Varshney, Y.B. Li, Talanta, 74 (2008) 518-525.

[60] B. Podola, M. Melkonian, J Appl Phycol, 17 (2005) 261-271.

[61] J.H. Lee, R.J. Mitchell, B.C. Kim, D.C. Cullen, M.B. Gu, Biosens Bioelectron, 21 (2005) 500-507.

[62] Y. Kuang, I. Biran, D.R. Walt, Analytical Chemistry, 76 (2004) 2902-2909.

[63] I. Biran, D.M. Rissin, E.Z. Ron, D.R. Walt, Anal Biochem, 315 (2003) 106-113.

[64] M.T. Giardi, L. Guzzella, P. Euzet, R. Rouillon, D. Esposito, Environmental Science & Technology, 39 (2005) 5378-5384.

[65] W.Y. Liao, C.H. Weng, G.B. Lee, T.C. Chou, Lab Chip, 6 (2006) 1362-1368.

[66] P. Ertl, S.R. Mikkelsen, Analytical Chemistry, 73 (2001) 4241-4248.

[67] N. Sanvicens, N. Pascual, M. Fernández-Argüelles, J. Adrián, J. Costa-Fernández, F. Sánchez-Baeza, A. Sanz-Medel, M.P. Marco, Analytical and Bioanalytical Chemistry, 399 (2011) 2755-2762.

[68] C.A. Rowe-Taitt, J.W. Hazzard, K.E. Hoffman, J.J. Cras, J.P. Golden, F.S. Ligler, Biosensors and Bioelectronics, 15 (2000) 579-589.

[69] K.E. Sapsford, C.R. Taitt, S. Fertig, M.H. Moore, M.E. Lassman, C.M. Maragos, L.C. Shriver-Lake, Biosensors and Bioelectronics, 21 (2006) 2298-2305.

[70] M.M. Ngundi, L.C. Shriver-Lake, M.H. Moore, M.E. Lassman, F.S. Ligler, C.R. Taitt, Analytical Chemistry, 77 (2004) 148-154.

[71] N.A. Rakow, K.S. Suslick, Nature, 406 (2006) 710-713.

[72] N.V. Kulagina, M.E. Lassman, F.S. Ligler, C.R. Taitt, Analytical Chemistry, 77 (2005) 6504-6508.

[73] C.A. Rowe, S.B. Scruggs, M.J. Feldstein, J.P. Golden, F.S. Ligler, Analytical Chemistry, 71 (1998) 433-439.

[74] C.A. Rowe, L.M. Tender, M.J. Feldstein, J.P. Golden, S.B. Scruggs, B.D. MacCraith, J.J. Cras, F.S. Ligler, Analytical Chemistry, 71 (1999) 3846-3852.

[75] K.E. Sapsford, A. Rasooly, C.R. Taitt, F.S. Ligler, Analytical Chemistry, 76 (2003) 433-440.

2

Environmental Monitoring Supported by the Regional Network Infrastructures

Elisa Benetti, Chiara Taddia and Gianluca Mazzini
Lepida SpA, Viale A. Moro 64, 40127 Bologna
Italy

1. Introduction

The aim of this chapter is the presentation of studies and research results concerning environmental monitoring techniques promoted by Lepida SpA across a wide area, the Italian Emilia-Romagna Region.

Lepida SpA *Lepida SpA* (2011) is an in house providing company established by a Regional Law (11/2004, "Regional Development of the Information Society") of Emilia-Romagna region, which consolidates a common vision and a collaborative approach with the local Public Administrations.

Lepida SpA was created in the end of 2007 by the Emilia-Romagna Regional Government, as unique shareholder and founder. Currently has 395 Public Administrations and Public Entities as shareholders. Lepida SpA is involved in the governance of the Regional ICT Plan which defines the regional ICT strategies and policies within the regional territory, acting as innovation facilitator among its partners.

The core business of Lepida SpA is represented by the regional ICT infrastructure but its operations range between telecommunication networks, digital divide and broadband networks strategies and ICT applications and services. Among the main activities and experiences pursued by Lepida SpA we can mention: the planning, development, management and monitoring of the telecommunications networks (fixed and mobile) of the P.A., including the deployment of new broadband networks (wired and wireless) within the region; the definition and implementation of suitable solutions for the Digital Divide topics and for the Next Generation Access Networks in order to ensure high speed internet for the citizens and businesses; the realization of ICT platforms and services for the Public Adminitrations (federation of authentication, payments, ..) that enable a large number of on-line services in favor of citizens and Enterprises; the realization of on-line services for e-Governement purposes and interaction between the P.A. and the Enterprises and citizens.

The infrastructure provided by Lepida and owned by the Public Administrations partners of Lepida spA, is an heterogeneous interconnected network covering the whole regional territory (more than twentytwo thousand square kilometers of area). It includes a regional area network (Optical Fiber) called *Lepida*, wireless networks (Hyperlan) that are extensions of *Lepida* which allow to solve Digital Divide in some mountain territories, and a regional emergency digital radio network (TETRA) called *ERretre*. A map of the Optical Fiber and Hiperlan link is illustrated in Figure 1.

The availability of this powerful infrastructure offers many opportunities for the P.A. to deploy and provide useful and interesting services to the citizens. Furthermore it represents a unique great regional test bed for the development and testing of new applications and services exploiting the potential of the ICT infrastructures.

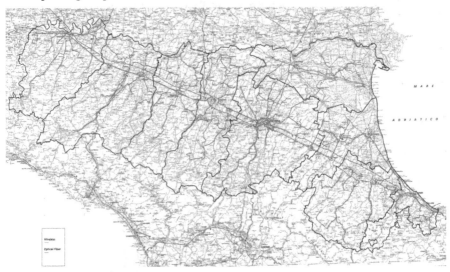

Fig. 1. Optical Fiber and Hiperlan link

In particular, this chapter will present efficient sensor network applications promoted by Lepida SpA and based on the regional hybrid access network, with the aim to realize environmental monitoring through an efficient usage of the territorial assets, by reaching therefore the important goal of public resources savings. The effort of Lepida SpA has been directed towards two primary directions: the first one is the exploitation of the Lepida SpA networks as a communication infrastructure that enables the messages exchanged by the softwares of data management that the Public Administrations already owns and uses for their environmental monitoring activities; the second one, besides the exploitation of the Lepida SpA networks like described in the first model, also proposes the usage by the Public Administrations partners of a proper software and/or hardware platform of data management, planned, tested and promoted by Lepida SpA.

In order to achieve this aim Lepida SpA has adopted a research method based on the following steps: 1) census of the sensor networks and communication networks used for environmental monitoring purpose, existent and operating across the whole regional territory 2) proposal of architectural, infrastructural and application service solutions 3) realization of experimental test-beds 4) adaptation and tuning of the solutions proposed during the second step in view of the results obtained during the third step 5) realization of a full service.

The census activity has been performed all over the Emilia-Romagna territory, by taking into consideration all the Public Organizations. This investigation has highlighted the presence of a huge amount of small sensor networks deployed all over the regional territory, consisting of spatially distributed devices for the monitoring of environmental conditions, such as temperature, sounds, pollutant, traffic, river and basin and also a lot of cameras for the video surveillance and video environmental monitoring. Typically they have been realized in the past as independent and autonomous systems, each one by using its own communication network to transport the collected data, each one by using its own sink to elaborate the data and each one belonging to a specific local Public Administration.

This scenario often brings the local Public Administrations to inefficient and expensive managements and maintenance of the data transmission, collection and elaboration. In such a scenario, the two working directions followed by Lepida SpA and mentioned above, can represent an effective way for the Public Administrations to pursue environmental monitoring activities while saving as much as possible resources and while following economies of scale. In particular Lepida SpA has defined a centralized architecture Taddia et al. (2009) based on a centre of collection, elaboration, management and diffusion of the sensor data that, by exploiting the hybrid access regional network, beside solving the inefficiencies can also provide further benefits that would be impossible to realize with independent and separate management systems. Let us mention just a few of the possible benefits enabled by the architecture promoted by Lepida SpA: data sharing among different Public Administration by saving the data property thanks to authentication and profilation solutions; correlation of data belonging to different Administrations. Lepida SpA has tested this architecture with some Public Administrations Taddia et al. (2010).

This chapter starts with a description of the adopted research method, by giving a comprehensive description of the first step of this research, the census of the resources available inside the Emilia-Romagna region. The rest of the chapter will describe more in detail how the aforementioned research method has been applied to three scenarios, by presenting three test bed actived by Lepida SpA in collaboration with three Public corporations: River Basin Consortium of the River Po affluents, Drainage Consortium of the western Romagna, River Monitoring for the Civil Protection of the Emilia Romagna Region. The three cases all exploit different network technologies among the ones offered by the the hybrid regional infrastructure, depending on factors such as the geographical position of the monitoring systems and the amount of data exchanged during the monitoring process.

2. Research methods

The method adopted by Lepida has performed, as a first step, an exhaustive census of all the automatic sensor networks deployed in the regional territory, not already integrated with regional sensor networks (sensor networks owned and managed by a regional Entity called ARPA *ARPA* (2011), Regional Agency Prevention and Environment for the Emilia-Romagna region). The Public Administrations in fact, may acquire and use their own networks in order to meet local needs that are within their competence. In order to carry out the census, all municipalities, provinces, the River Basin Consortium of all the provinces and the civil protection have been contacted. For each network, the following items have been surveyed: type of measured data, number of sensors used, number of data loggers used, transmission media and the Administration involved. Offices for environment and mobility, farming, civil protection and provincial police, have been consulted in main cities of each province. Received responses have been inserted in a database containing the following information: the owner Administration, the service manager, the operator, type of monitoring, number of stations installed, number and type of sensor used and the transmission media. Subsequently an analysis of these responses has highlighted different trends and consolidated needs, depending on the responsible Administration and its skills and jurisdiction. Various types of networks, used by different Administrations, that have been found thanks to the census, are shown in Figure 2.

Types of Monitoring Systems	Entity					
Type of Monitoring	Municipality	Province	River Basin Consortium	Mountain Association	Regional Civil Protection	Total
Videosurveillance	14			1		15
Accesses control	8					8
Traffic Monitoring	7	1				8
Landslides Monitoring			1	6		7
River Monitoring	1	4			1	6
Metereological station	1	4				5
Traffic light infraction	4					4
Messages	3					3
Traffic light optimization	3					3
Speed control	1	1				2
Acoustic	1					1
Bycicles count	1					1
Parking control	1					1
water monitoring			1			1
Building stability	1					1
Total	46	12	6	1	1	66

Fig. 2. Types of monitoring systems related to different entities

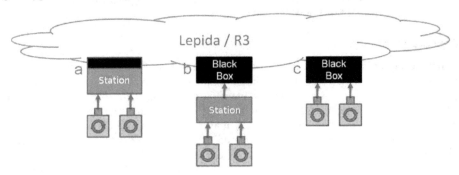

Fig. 3. Models of integration.

Afterwards, for each type of monitoring system, the type and number of sensors used have been mapped, so that their spread could be better understood. As a result was noted that the most common sensors are: the inductive coil (its low cost and its simplicity of use have made it the leader in sensor networks for traffic monitoring); the camera (used by local Public Administrations in response to a need of an improved security for citizens, furthermore the wealth of information intrinsic in its data detected, that is a stream of images, makes this sensor suitable also for other applications such as traffic monitoring or rivers flow control); the inclinometer (its purpose is related to applications for landslides monitoring).

A further analysis about possible efficient architectures that could be proposed to shareholders Administrations, pointed out that is desirable to integrate all existing networks, both for surveillance systems, which are increasingly spreading throughout the territory, and for landslides monitoring, currently managed in a summary way. The presence of a unique wide regional network on the territory, composed by Lepida and *ERretre*, makes this integration possible and it represents also the opportunity to have a uniform and guaranteed transmission of data gathered by all sensor networks. Three different models of integration with Lepida network have been proposed, as shown in Figure 3. Two of them exploit a small hardware and software module programmed by Lepida SpA and called BlackBox, which is mainly devoted to the integration between the communication infrastructure and the sensors.

(a) IP and TETRA driver: a monitoring station, provided by third-parties, on one hand interfaces to sensors and on the other hand to the most suitable telematic infrastructure, chosen between *Lepida* and *ERretre*, through suitable management drivers;

(b) Gateway: a control board interfaces to the monitoring station provided by third-parties through a proprietary protocol or through the standard protocol Modbus. The BlackBox, on the transport network side, provides the most suitable driver depending on the transmission media that will be chosen;

(c) Direct interface: the BlackBox could directly interface to sensors and at the network side performs the gateway functionalities as described in step (b).

The results obtained by the census activities have given the room of defining a suitable architecture able to face the problematic arisen, both in terms of data management system and in terms of communication technologies and infrastructures. Starting from this architectural solutions, some test-beds have been activated nad they will be described in detail on the following Sections.

3. River basin consortium

The subject involved in this testing is the River Basin Consortium of the "Po" River, an agency that deals with the emergency activities related to the water channels and seismic events of "Piacenza", "Parma", "Reggio-Emilia" and "Modena" territories.

The current sensor network that the River Basin Consortium owns and uses presents a lot of problematic aspects: these are particularly correlated to the communication networks currently used, and to the management and storage of data. The data management and storage are fully delegated to private companies that do not offer a system able to ensure the necessary levels of availability and persistence of data. Furthermore, data are distributed on different servers that differ in technology and data representation: there is not a single centralized system that could gather all available information in a standardized format.

Lepida SpA in this case has proposed to the River Basin Consortium of the "Po" River a test-bed activity based both on an interface to the communication infrastructure provided by the *ERretre* network, and on a prototype of a data management center that could satisfy all the needs requested by a full monitoring system.

3.1 BlackBox

The BlackBox prototype has been realized through a control board based on ARM Linux. As shown in the second model of Figure 2 it could be connected transparently to all proprietary tracking stations which export the Modbus interface. This is an open serial communication protocol, master-slave or master-multislave, developed to transmit information between several PLCs (Programmable Logic Controllers) through a network connection and has become, over the years, a de facto standard communication protocol for the industry. Otherwise, in the third model schematized in Figure 2, the BlackBox provides the management of three different types of sensors: digital sensors, that could also be connected in a multiple modality through a multi-master and multi-slave communication bus; a single generic alarm button; a single serial sensor.

In order to properly handle these three types of sensors, for each one of them a dedicated parallel task has been implemented in the BlackBox: this ensures the management of any kind of warning, even asynchronous, from sensors. Furthermore, the BlackBox interfaces to the network both to transmit data and receive commands, through two different ways: either using the Ethernet connection for communication via IP or the serial connection for communication via Tetra terminal, in this case by SDS. The software is based on a task that periodically requests a measure to all the sensors connected and sends them to the data collection center, also managing the reception of any command configuration parameter, such as changing the sampling rate or actuating connected devices, for example an acoustic or light signal. A software unit receives as input the messages sent by the BlackBox, interpreting and storing them properly. The server where this unit resides, is interfaced both to the IP network and ERretre through a modem connected to a ttyUSB port. In particular, when a message is received the unit, according to the opcode message and to the sender sensor typology, properly extracts the information and stores them in a table or in another textual file available in the system and used by the entity, considering them as a single sensor in a unique instant of sampling. A single message, in fact, could also contain several measures of a unique sensor but related to subsequent sampling instants, or measures sent by different sensors but related to the same sampling instant.

The experimentation with the River Basin Consortium is based on the second model of integration and, due to the isolated location of the test-bed site which does not allow an ethernet connection to the Lepida Network, the communication is done via SDS.

3.2 Landslides monitoring

The test-bed organized by Lepida SpA was installed on the 16th of July, 2010, at the landslide by Fosso Moranda, in the Polinago municipality, province of Modena. It consists of a proprietary survey station (Datalogger) with two biaxial inclinometers at different depths, which perform accurate measures related to millimetric movements of the ground, and a piezometer, which measures the hydrostatic pressure, attached to it. The BlackBox is connected to a Tetra modem for the transmission of data, according to the configuration where the detection station acts as a slave and the BlackBox is both the master and the gateway towards the Tetra transmission network, as shown in Figure 4. The system is powered by a photovoltaic panel and is normally turned off. At a scheduled sampling rate, typically every hour, the monitoring station will "wake up" and control the power supply of the entire system: both Tetra modem and Blackbox. The BlackBox requests to the station data from sensors, then sends the response message to the data management center and commands the proprietary station, that supervises the power control, to shut down the system. The communications between the proprietary station and the BlackBox physically occur through a serial connection and logically exporting at both sides the standard interface Modbus, as previously explained. In addiction to specific parameters the system also includes the monitoring of the backup battery level, which is useful in checking the functioning of the whole automated measurement system. All processed data have a low weight, that is about

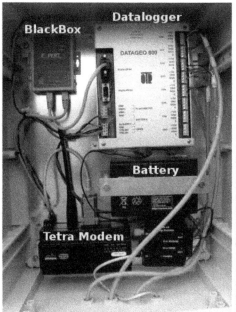

Fig. 4. Cabinet and installation site

20 bytes for each transmission. However the test-bed is highly significant because it is related to a real installation site characterized by particularly hostile conditions, located in an isolated area without any continuous electrical power available. The activation of the whole system has been made possible thanks to a survey about Tetra modems on the market and the identification of which one of them are compatible with the regional network. These could be, unlike ordinary terminals, turned on and off through a simple contact, providing less current absorption and having a lower price.

As a consequence of the good results achieved, the River Basin Consortium and Lepida Spa has arranged a second experimentation phase that should include three new installations connected to multiple sensors and an extension of the BlackBox features, such as remote log retrieval, remote change of the frequency sampling.

3.3 LabICT and Data Management Center

In a previous research phase, a prototype of a unified Data Management Center (DMC) was internally carried out at Lepida SpA R&D Laboratory, in order to receive data, normalize and validate them depending on operation thresholds according to their type and brand. A further analysis of data also allowed a cross-checking of different sensors to trigger alarms for values exceeding from defined thresholds, or for failures. An initial authentication foresaw a base profiling that determined primarily two types of users: basic and operator. for the basic one, thanks to a web interface, a real-time graph with the last samples gathered could be visualized, an historic archive including all measurement done could be consulted and these values could be sent, in a graphic format or through a pdf table, to an e-mail address. Moreover a map showed the location of the stations and the BlackBoxes installed all over the regional territory; for the operator one, in addition to the basic features, this type of user could insert new units and sensors pertaining to his entity or his partners. Finally he could define new alarm thresholds.

Although this system was quite complete, it had been implemented with the aim to show its potential in environmental monitoring and some features were in an embryonic stage of development. As a result of an increasing interest and a great satisfaction showed by the entities, at the end of the experimental phase, starting from this previous experience the prototype is evolved into a more complex and efficient solution taking advantage of the LabICT-PA (Laboratory for Information and Communication Technology for Public Administration). The LabICT-PA, created in 2007 by the Emilia Romagna region, is part of the Regional High Technology Network and aims to accelerate innovation in public administration. Since 2011 LabICT-PA is also a member of Europeean Network of Living Labs *ENoLL* (2011). The organizational model of LabICT-PA is based on the living labs, where the functional requirements and specifications are defined by and with the users, that is Public Administrations. Design and testing phases will be also carried out through a continuous dialogue with end users. The main partners and their roles in this living lab are: the Emilia Romagna Regional Government that determines the police through the ICT plan; Lepida SpA that, as in house providing company established by a regional law, coordinates activities and provides technical competences and effort; almost 400 public shareholders of Lepida SpA that represent end users; almost 100 business partners, called the club of stakeholders Lepida, that are the think tanks that create added value for PA and for the market; finally universities and research institutes serve as research partners for the laboratory. In this sensor networks context, LabICt-PA has created a fully working prototype, non-engineered, of data management center for sensor networks.

3.3.1 Architecture

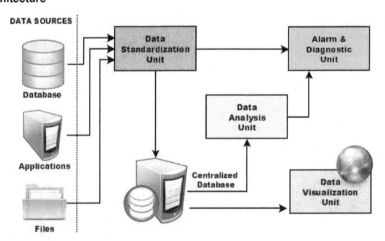

Fig. 5. Data Management Center Architecture

This project aims to integrate all sensor networks deployed in the region through the implementation of a shared platform that could uniformly handle all kinds of environmental data. Firstly, the database of the previous prototype was completely revised to improve the management of the data, intended as a single measure detected by the sensor, making it the most generalized as possible. In fact, the main architectural features are:

- Modularity: each block is independent and communicates through the exchange of XML files;

- Scalability: each module can be implemented on different physical machines;

- Configurability: main operating parameters could be defined in a database, including the definition of new types of sensors, thresholds, alarms, and so on.

All sensors have been schematized in a hierarchical way so that multiple sensors may depend, whether or not, on a BlackBox, which can be connected to another unit, too, for example proprietary stations. Each one of these elements is categorized as a sensor, this is because they are all able to send and receive signals, moreover each sensor can perform different types of measurement with different timing for the acquisition. Finally, measures may be punctual, aggregated or their avarage is calculated, depending on various time intervals. In addition to the tables dedicated to sensors management, the database also includes additional tables necessary to provide addresses, ticketing, alarms, profiling, logging. The Middleware, the Control Center and the Monitoring Center consist of opensource units (Figure 5): each one has its own characteristics, in order to satisfy all the features proposed and also maintain a huge flexibility, in fact each unit inside t he project is independent from the others. The whole managing of data within the Data Management Centre can be divided into three main phases, acquisition, processing, viewing, and this allows to describe each single functional unit. Heterogeneous data sources will be homogenized by the first standardizing unit and then the measures will be evaluated by the analysis unit that will validate them and will check all alarm thresholds. The alarm and diagnostic unit will be contacted by both units and manages and logs the events. Finally, the validated data will be displayed by the visualization unit through a web interface. Communications between two different units are done by using Web Services.

3.3.2 Data standardization unit

This module is the interconnection and standardization middleware between the data and the central unit, therefore plays the role of collector and uniforms data sent from different sources storing them in the database of the DMC. It is based on the following elements:

- Atomic modules for data retrieval: are used to retrieve the data, both automatically at a preset timeslot and on-demand, gathering data from various sources or databases. Inside each atomic unit the access procedure and the detailed commands used to retrieve data from a specific source are specified.

- Atomic units manager: is always active and coordinates the required units. It also serves as a collector for messages sent by the individual atomic modules and redirects them through the units of communication, alarming, diagnostics and data analysis.

- Communication unit: it allows the manager to communicate with other modules inside the platform, on one hand by collecting the total number of messages and errors from the manager, on the other hand receiving as input all requests sent by the DMC and directed to the manager.

Output messages produced by this unit are: the standardized data subsequently stored on a centralized database, the notification messages that new data has been inserted in the database so that the proper unit could start to analyze them, errors and log messages that are transmitted to the diagnostic unit.

3.3.3 Data analysis unit

Its purpose is to control the last data processed by the unit of standardization and to do periodic monitoring on the centralized database in order to trigger the following types of alarms:

- Failures: they occur in two cases, when the unit detects that a certain sensor does not send values in a timeslot that is longer than the sampling rate specified for that sensor, or when the measure is not performed correctly in respect to the working range of the sensor.

- Alerts: several alert situations can be assigned to a unique measure and they may depend on the overcoming of a minimum or maximum threshold, or on an excessive increase or decrease of the measure compared to the previous value stored. The amount of subsequent occurrences of the same state of alert, that must be verified before triggering the proper signaling to the unit of alarms and diagnostics, could be also specified.

- Simplex: this event is triggered as a result of the simultaneous testing of multiple alarm conditions. In a unique simplex both alerts and failures could be associated, linked together by logical operators (and, or, not) so that an event could be characterized by critical conditions based on multiple sensors in very complex relationships.

When one of these alarms occurs, it is communicated to the alarm and diagnostic unit specifying which sensor has triggered the alarm event, the type of event and which alert message has been associated to the event, so that all information needed are forwarded to the dedicated unit, due to simplify and speed up its alarm procedures.

3.3.4 Alarm and diagnostic unit

In addition to alarms generated by the analysis unit, all units part of the system architecture could send error messages in case there is a generic malfunctioning in the DMC such as database connection errors, query failed, units that are not working and so on. The diagnostic unit is implemented using a web service SOAP and handles all the incoming XML requests storing and logging them properly. If they are associated with one or more alarm procedures the unit sends the warning message to one or more users by an email, an SMS or an SDS on a Tetra terminal. Finally, the unit manages generic events that could be scheduled at certain timeslots and which may be linked to the linear chart of a sensor so that when a value exceeds from its alarm, an e-mail should be sent not only including a warning message but also with the graph related to the sensor involved as attachment, due to have a visual feedback of the current situation.

3.3.5 Data visualization unit

This unit is based on a web site consisting of several forms that allow the user to query and monitor the various data structures included into the DMC. All the forms have been integrated into a single portal and are made up of different tabs, available on the main screen of the site. A tree view in the left side of the web site represents all the system control stations and sensors connected to them, then each sensor will match one or more type of measures. This tree is generated according to the initial login: in fact an association is possible between a profile and a user, that specifies which sensors he could visualize. The icons of the tree have different colours to provide visual indications about the status of each sensor: green if the sensor works correctly, red in case of alert, yellow in case of failure and gray if is disabled. The tree view allows the selection of multiple components. A geo-referenced map of the region is also provided in the homepage and the markers shown on it indicate the stations installed

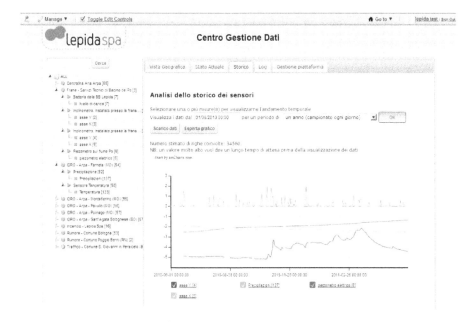

Fig. 6. Data Management Center homepage

using colors in agreement with those defined for the tree view icons. Clicking on a marker a description of the unit and a description of the sensors connected to it are shown. The additional tabs are:

- Real-time monitoring: it provides a graphical and tabular representation of the last data sent by the sensors. The measures to be displayed can be selected through the tree view. The chart adapts its time scale according to a selection done in a drop down menu and then automatically updates itself every 5 seconds. In Figure 6, for example, a multiple real-time chart related to one inclinometer, the piezometer and one ARPA pluviometer is shown.

- Analysis of historical data: in this tab, data could be analyzed with an historical depth that is greater than the one on the real time tab, selecting a start date and a period to display. It 'can be downloaded locally both in a graphic and a tabular format.

- Logs viewing: provides a list in chronological order of all the significant events detected in relation to sensors failures (started or stopped), alerts (started or stopped), invalid values, and so on.

- Platform management: supplies some statistics about the current state of the system, for example the status of the various units involved and an overview of all detected events.

4. Drainage consortium of western Romagna

A Drainage Consortium is a public corporation that coordinates both public actions and private activites concerning the drainage of its territory of scope. For example, hydraulic

Fig. 7. Drainage Consortiums in Emilia-Romagna region

security, management of the waters intended to the irrigation, involvement into urban planning, environmental and agricultural heritage protection can be considered typical activites and actions covered by a Drainage Consortium.

In Emilia-Romagna region eight Drainage Consortiums exist, subdivided depending on their area of scope, as illustrated in Figure 7. All of them are partners of Lepida SpA, therefore Lepida SpA is legitimized to be involved for the support of their activities, by favouring economies of scale.

Currently each Consortium manages a suitable small sensor network, consisting of a set of data logger, devoted to hydrographical detection and remote control functions, thanks to the use of Programmable Logic Controllers (PLCs) and sensors connected to the data loggers. Furthermore each Consortium has got a suitable monitoring system (typically a server hosting a software system of data management) devoted to the collection of all the gathered data. Data are exchanged between data logger and server and among the data loggers (often there is the need to spread some specific control command from a data logger to other data loggers, by following as a sort of tree communication path) by using analog or GSM technologies (generally GSM is used to send alarm messages to people that need to be activated in case of danger or alarm situations while analog communication channels are used for the data collected by the sensors). Economies of scale could be found in such a scenario, by exploiting the network infrastructures owned by Lepida SpA.

For this purpose Lepida SpA will support the Consortiums, by starting from the Drainage Consortium of the Western Romagna *Lugo* (2011), which has been involved in a test-bed stage. The condition of the equipement managed by the Drainage Consortium of the Western Romagna, before the mentioned test bed stage, can be summerized as follows. It is composed by fifteen data loggers, each one including a PLC with some sensors for the hydraulic data collection and an analog communication module. Each module communicates the monitored data trough UHF channel while the alarm signals are sent through GSM network, by means of Short Message Service (SMS). The monitoring activity is mainly performed by following a polling communication protocol: a central server, devoted to the data collection and elaboration, polls each data logger every thirty minutes, by receiving the data that the sensors connected to the data logger have recorded at a one minute frequency during the last thirty minutes; every poll requires about 300 bytes for each data logger. Besides the polling scheduled activity, the central monitoring system has also the possibility of directly poll a specific data logger at any instant, for example in case of emergency; in this case the polled data logger will send the data collected since the last polling.

The effort of Lepida SpA has been addressed to the communication technology used by the described system. In particular by offering the opportunity of exploiting the regional tetra radio infrastructure of the *ERretre* network as a data transportation driver. This represents for the Consortium an opportunity of economic and resource saving, by replacing the old analog radio with the modern digital tetra radio. Furthermore the *ERretre* network can offer other fundamental advantages: it can guarantee a full coverage inside the whole regional territory, allowing also intercommunications among the devices of different Drainage Consortiums in Emilia-Romagna region; it can offer guaranteed communications, avoiding the congestion events that instead may occur in a GSM network. Another role of Lepida will be the support in defining the technical specifics for a future furnishing of digital tetra radio to be used in a long-term solution when the test-bed stage will be definitely ended.

The test bed activated in collaboration with the Drainage Consortium of the Western Romagna exploits the *ERretre* network, with single slot packet data communication policy.

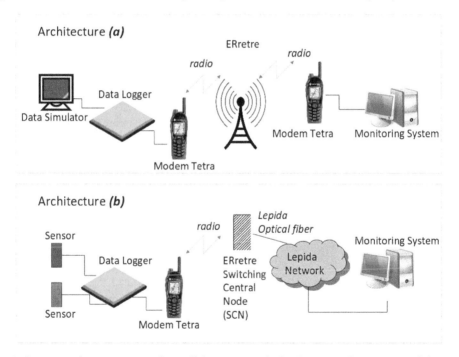

Fig. 8. System architectures tested in collaboration with the Drainage Consortium of the Western Romagna

The test bed has been conducted through two phases. The first one has been realized at Lepida Spa R&D Laboratories, by implementing two different architectures, illustrated in Figures 8, which leads to two different network performances. In architecture *(a)* both the data logger and the monitoring system are connected to two tetra radio modem: in this case any data tramission requires the activation of two radio links (the first one between the data logger and the *ERretre* base station covering the area in which the data logger is located and the second one between the monitoring system and the *ERretre* base station covering the area in which the monitoring system is located). A more efficient usage of the network resources is presented by architecture *(b)*, that exploits the existent cabled connection (optical fiber) between the *ERretre* Switching Central Node and the Lepida network infrastructure. The main advantage of this solution is represented by the reduction of the traffic routed inside the *ERretre* network, which institutional role is mainly represented by emergency communications, and the exploitation of the bandwith offered by the optical fiber link. concerning the data logger element, this has been emulated during the tests realized at the Lepida Laboratories, with a specific data simulator connected to the PLC of a real data logger. Both the two architectures have shown very good results when tested at the laboratories. These are both fundamental: one optimizing the network performances, the other enabling the implementation of sequential commands to more data loggers.

The second experimental phase has been activated with a real data logger, collecting real data from the sensors connected to it. The remote data logger was located at "Mordano" city and the central monitoring system was located at"Lugo" city, each one equipped with a tetra radio. The network architecture implemented is the *(a)* type (by referring to Figure 8 *(a)*, the only difference in terms of the real sensors that were used instead of the data simulator appliance), mainly because of two reasons: the head office of the Drainage Consortium of Western Romagna, located at "Lugo", was not yet connected to the Lepida optical fiber; the Drainage Consortium needs to activate also communications among each single data logger, by following a sort of tree path, in order to allow a data logger to send particular commands directly to other data loggers (for example a data logger being able to sequentially command a more draining installations).

Basically the communication is realized by following a polling communication protocol: the central server polls the data logger every thirty minutes, by receiving the data that the sensors connected to the data logger have recorded at a one minute frequency during the last thirty minutes; every poll requires about 300 bytes. Nevertheless the communication is bidirectional: the PLC is devoted both to the data gathering from sensor like temperature, water levels and water flow and to the reception of commands from the central server, like bulkheads closing or pumps activations; similarly the server receives the data gathered by the sensors and it can send remote commands to the PLC.

Figure 9 shows the place where the draining pump of "Mordano" is located, the control systems with PLC and data loggers and the tetra modem used for the test bed. Figure 10 shows the panel observable at "Lugo" with the monitoring system. The data management software and the data logger have been implemented and supplied by private enterprises that directly collaborate with the Drainage Consortium of Western Romagna.

This experimental phase has shown optimal results in terms of network performance and communications reliability. The test bed will be soon extended to fifteen other PLCs. The long-term installation will see the architecture *(b)* for the connection of the monitoring system and the architecture *(a)* among the single data loggers.

5. Civil protection

A further test bed has been created with the aim of realizing integrated systems for data and video communication on the Emilia-Romagna Regional territory, particularly for hydro geological and hydraulic risk, in cooperation with the Civil Protection of the Emilia-Romagna

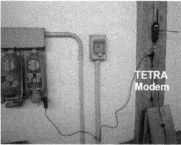

Fig. 9. Draining pump located at "Mordano" city.

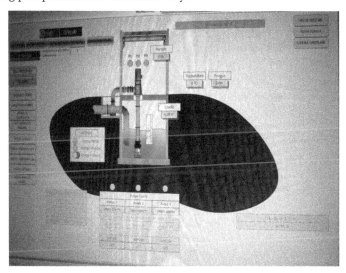

Fig. 10. Monitoring systema located at "Lugo" city.

region. Generally speaking, Civil Protection works around the whole national territory and it is divided in regional agencies, which are subdivided in smaller agencies, devoted to the execution of pregressively more local aspects. Its main working acivities are forecasting and preventions, rescue activities, emergency trainings.

The project, promoted by the Civil Protection of the Emilia-Romagna region, aims to the realization, with the technologies at disposal nowadays, of a set of integrated systems devoted essentially to the monitoring and to the communication of data/video over the Emilia-Romagna Regional territory, to fulfill the requirements of the Civil Protection: prevention and management of hydrogeological-hydraulic risk, emergencies, disasters, trainings. More precisely, the project includes the following items:

- system for the visualization of large area and for the visual integration of images and data, both newly generated than preeexistant, trough a rear projection videowall located at the Operating Center of the Civil Protection of the Emilia-Romagna Region, in Bologna City;

- capturing systems of territorial images for the permanent monitoring of critical area such as bridges or other infrastructure located near rivers or torrents or dry detention basins;

- capturing systems of territorial images for emergency or temporary situations, devoted to the monitoring of rivers or landslides or disasters area or even training activities, by exploiting a transportable device and a motor vehicle.

The project has started with a first stage implementation that has been developed and employed as an experimental test bed, usefull to understand the limits and strength of the architecture of the whole system. More in detail, this first stage has realized the videowall and a sort of "mini network" of video monitoring, which specfications are explained as follows:

- One video camera and related data/video transmission systems (in Hiperlan technologies at 5.4GHz) devoted to the permanent video monitoring of the river "Savio" near "Cesena" city and located at the railways bridge over this river 11.

- One kit of video monitoring and recording and related data/video multistandard transmission systems (Hiperlan, TETRA, UMTS, Ethernet, etc...). This kit has been installed over a transoprtable device (a trailer truck) devoted to temporary video monitoring activities in area such as rivers, torrents, landslides or other situations of emergency or temporary training exercises.

- One kit of video monitoring and data/video transmission installaed over a motor vehicle (in this case a vehicle owned by the Group of Amateur Radio Operators Volunteers of the city of "Imola" was used) devoted to monitoring activities during training exercises or disasters or emergency situations.

- One hardware and software centralization system (server) for the centralized recording and management of the data/video collected by the remote video systems listed in the previous items; the mentioned server was located at the Data Elaboration Center of the Emilia-Romagna Region, in "Bologna" city;

- Four workstations related to the four systems listed above: two located at the Operating Center of the Civil Protection of the Emilia-Romagna Region, together with the videowall; one located at the of Operating Center of the River Basin Consortium of the city of "Forlí"; the last one located at the Operating Center of the River Basin Consortium of the city of "Cesena".

Lepida SpA has been involved in this first stage of the project for the definition of network architectural specifics and for the configuration of the network active devices involved (switches, routers...), besides offering to the Civil Protection the usage of its network

Fig. 11. Design of camera installation at the railways bridge over the "Savio" river, near "Cesena" city.

infrastructures. A full description of the architecture realized for this first stage test-bed is illustrated in Figure 12. More precisely, Lepida SpA fiber network has been used as a backbone between the camera and the centralized server located at Operating Center of the Civil Protection of the Emilia-Romagna Region, while the connection between each single camera and the closest point of presence of Lepida network has been designed and implemented as an ad hoc wireless link, with the proper and better technology (Hiperlan, TETRA...), depending on the specific use case (temporary or permanent monitoring station).

For this reason, this stage has represented for Lepida SpA mainly a test bed in which testing the network performances serving as collector of many data/video distributed all over the regional territory. The cameras used for the test bed stage are professional Megapixel IP cameras supporting H.264 video streaming. As far as the centralized system of data recording and management is concerned, the first stage used a solution based on a commercial product (Genetec Omnicast).

The item monitored by the camera is the "Savio" river. The amount of traffic over the network is about one image of thirty kBytes every fifteen minutes. When the TETRA channel is used, each image is sent by exploiting the single slot packet data type of communication, that offers a 3kbits/sec of bandwith. The amount of traffic exchanged is compliant with the performances offered by the TETRA channel.

The first stage test bed has shown positive and promising results, leading to the definition of guideline for a second stage of this project. This second stage will involve the realization of at least thirty other video camera located at different river area already defined and spread all over the Emilia-Romagna regional territory. The main requirement of the future new installation is the full compatibility with the infrastructures, the hardware and software equipments already used during the first stage of the project.

5.1 Video management center

During the test bed realized in collaboration with the Civil Protection, the role of Lepida SpA has been mainly focused over the definition of the proper architecture of the communication

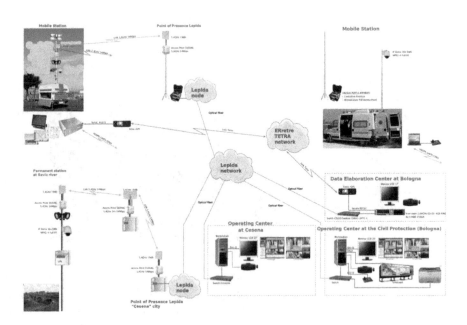

Fig. 12. Architecture of the first stage of the test bed realized in collaboration with the Civil Protection.

network to be used, in terms of the selection of the proper communication driver technology (hiperlan, TETRA, fiber...) and in terms of the proper addressing policy to be adopted among the different Entities (Regional Administrations and Regional Data Elaboration Center, Civil Protections...) involved. During the first stage test bed, differently than during the activity with the River Basin Consortium, no central data management system has been proposed and tested by Lepida SpA, since a commercial one was currently at disposal of the Civil Protection, at least for the purpose of the first stage activity.

As the census results has shown, a lot of Public Administrations in the Emilia-Romagna region are equipped with videosurveillance cameras for the purpose of public security maintenace. As the census activity has highlighted, these video show the same behavior of the environmental sensors installations (temperature, landslides, pollutions...): they often have been installed in former times by each single Administration, as separate and independent systems, resulting in a lot of camera managed by autonomous, independent and very expensive videosurveillance appliances. Furthermore, during the more recent years, expecially the smaller Public Administrations located in the Emilia-Romagna region, have expressed the necessity of providing themeselves with videocamera systems, asking support to Lepida SpA, the natural and institutional reference for the resource efficient development of their activities. More precisely these Administrations needs support concerning the definition of the technical specifics to include in public announcements devoted to the cameras, ad hoc communication infrastructures and video management systems. Such a scenario, like the scenario explored with the River Basin Consortium, points out the necessity of a central management system that could help resource saving policies in the management activities correlated to the video data. Lepida SpA, in order to readily face and solve the necessities that arise from the regional territory has designed a global architecture for an efficient management of all the video data that potentially could be generated by the cameras and videosurveillance systems installed by all the Public Administrations in Emilia-Romagna region.

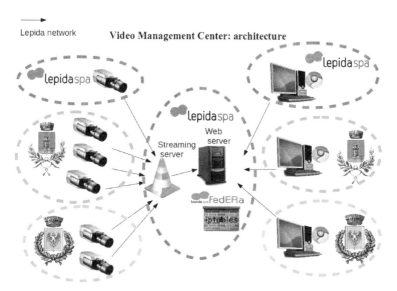

Fig. 13. Architecture for the management of camera strems produced by the Public Administrations.

The architecture proposed by Lepida SpA is illustrated in the Figure 13: Lepida SpA will provide a central server (or a proper cluster of servers) devoted to the storage of all the video streams recorded by the cameras, that will be collected by exploiting the Lepida network infrastructures; this centralized system will host also a full video management service, completely designed and implemented by Lepida SpA with the exploitation of open source technologies, that thanks to the Lepida network infrastructures, will offer to the remote Public Administrations client workstations features like: live streaming view, recorded video of at least one week old (in accordance to the italian law), downloading of recorded video.
Lepida Spa has implemented these features in a Video Management Center prototype. Figures 14 and 15 show some snapshot. More precisely this prototype is composed by a streaming server that collects all the streming video produced by the cameras and streams them to a management server. The management server functions as a web broadcaster of the data streamed by the streaming server, a choice that allows a big scalability in terms of network traffic: each client can watch a live video directly from this streaming server, avoiding the multiplication of network traffic that would be necessary if each client would refere directly to each camera stream. Management and streming server can be hosted in the same hardware server (specifically, the prototype realized follows this policy). The management server is implemented with web services technologies (php, apache, html, javascript, ajax) and it provides a web interface with tabs for the live view, for the recorded video view and for the downloading of the recorded video. In order to be compliant with all the requirements contemplated by the italian law in terms of privacy related to the management of video

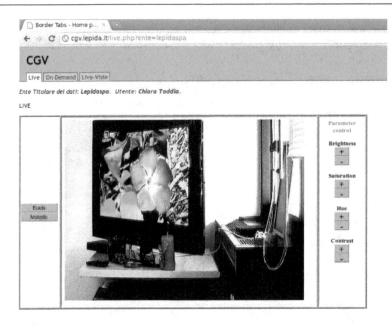

Fig. 14. Video Management Center prototype (CGV): streaming live view feature available on the web interface. The camera shown and listed are used in Lepida SpA R&D laboratories as a test-bed.

data, the prototype also implements other features such as: strong authentication of all the users that can access the web interface; profilation of the users that can access the web interface, in order to limit and specifically select the features each users is enabled to access; logs collection related to all the operations that each user makes on the web interface. The authentication is performed by exploiting the "fedERa" system. "FedERa" *fedERa* (2011) is the regional authentication system designed and promoted by Lepida SpA; it is a federated authentication system that allows the access to all the online services offered by the federated Public Administrations, with the usage of a single username-password that is valid for all the federated services. The profilation system has been implemented ad hoc for this prototype and it allows to differentiate the features that are available to each user: live view, view on demand of the recorded video, download of the recorded video. The profilation system, strictly correlated to the authentication system, allows the creation of logs to trace the activities of video data managements, as required by the italian privacy laws: the log systems collects information about the timestamp, identity of the user and specific operation made on the Video Management Centre (live view, on demand view, download). Actually the Video Management Center prototype has been tested by Lepida SpA internally. In a couple of months it is scheduled the start of a test bed usage of such a system in collaboration with a small Public Administration of the Emilia-Romagna region, for public security purpose. The architecture designed for security videosurveillance purpose and the developed prototype could nevertheless also be used in scenarios such as the one described by the project of the Civil Protection, so for environmental monitoring purposes that are based on video images.

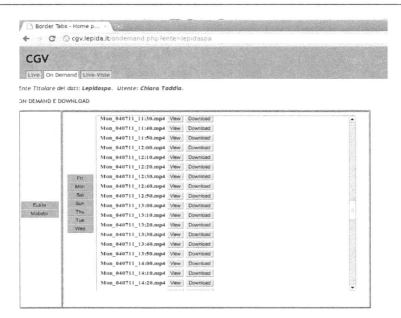

Fig. 15. Video Management Center prototype (CGV): video recorded on demand feature available on the web interface. The camera shown and listed are used in Lepida SpA R&D laboratories as a test-bed.

6. Conclusions

This publication has highlighted some interesting aspects of environmental monitoring mainly related to the Public Administrations context. Let us stress that, expecially in the Public context, an effective usage of the available resources and a development based on economies of scale are fundamental. The research method proposed and adopted by Lepida SpA has represented, as the described test-beds have shown, a valid instrument for the efficient planning, without any resource wasting, both phisical and economic, of all the environmental monitoring activities managed at a regional level. In particular the method proposed has given the opportunity of increasing the knowledge of all the infrastrucutures and system concerning the environmental monitoring field, active and used inside the whole regional territories, resulting in a more aware development of new hardware and software solutions able to face with the practical problems and necessity raised by the Public Administrations of the Emilia-Romagna region. The test-beds activated and described in this chapter will offer in the next months the chance to properly set and define all the systems variables and parameters in order to become useful models for the definition of future full services devoted to the Public Administrations.

7. Acknowledgment

This chapter was prepared in collaboration with Eng. Stefania Nanni, Laboratories ICT Manager at Lepida SpA, Eng. Anna Lisa Minghetti, Lepida network Manager at Lepida SpA, Eng. Federico Marcheselli, ERretre network Manager at Lepida SpA, Eng. Fabio Brunelli, Services Development area at Lepida SpA.

Authors would like to thank the following people and organizations, for their competence and their data: River Basin Consortium of the "Po" River, Civil Protection of Emilia-Romagna Region, Drainage Consortium of the Western Romagna, Marconi Labs, Dab, Misa, Eltel4, IEIIT-CNR Bologna, IConsulting, MEEO.

8. References

ARPA (2011).
      URL: *http://www.arpa.emr.it*
ENoLL (2011).
      URL: *http://www.openlivinglabs.eu/*
fedERa (2011).
      URL: *http://www.lepida.it/lepida-per-attivita/servizi/autenticazione-federata-federa*
Lepida SpA (2011).
      URL: *www.lepida.it*
Lugo (2011).
      URL: *http://www.bonificalugo.it*
Taddia, C., Nanni, S. & Mazzini, G. (2009). Technology integration for the services offered by the public administrations, *IARIA Neutral, Cannes, France* .
Taddia, C., Salbaroli, E., Benetti, E. & Mazzini, G. (2010). Centralized management of data collection over hybrid networks, *IEEE ACCESS, Valencia, Spain* .

3

Monitoring Information Systems to Support Adaptive Water Management

Raffaele Giordano, Giuseppe Passarella and Emanuele Barca
Water Research Institute - National Research Council, Bari, Italy

1. Introduction

Decision making in water resources management is widely acknowledged in literature to be a rational process, based on appropriate information and modeling results. Information plays a fundamental role in improving our understanding of the consequences of, and trade-off among, the alternatives in water resources management.

Environmental monitoring networks have the potential to provide a great deal of information for environmental decision processes. Monitoring is widely used to increase our knowledge both of the state of the environment and of socio-economic conditions. Environmental monitoring has demonstrated its capacity within resource management to support decision processes providing knowledge of baseline conditions, to detect change, to establish historical status and trends, to promote long-term understanding or prediction, and to establish the need for, or success of, interventions.

Our knowledge of the complexity of water system processes is increasing, together with our awareness of the uncertainty and unpredictability of the effects of water management on system dynamics. Consequently, the demand for environmental information is growing posing new challenges to monitoring system design. This chapter discusses these new challenges and proposes an innovative monitoring design approach to deal with complexity. The conceptual architecture of an Adaptive Monitoring Information System (AMIS) is proposed. The AMIS properties are used in this work to define a framework to assess the capabilities of current monitoring systems to support water managers to cope with complexity and uncertainty. The framework is used to identify the main limitations and to define the potential improvements of TIZIANO monitoring system, developed to monitor the state of groundwater monitoring in the Apulia Region (South Italy).

2. New challenges for monitoring systems and information management in Adaptive Management (AM)

Incorporating uncertainties about future pressures on river basins into water resources management sets new challenges for environmental resources management. One learning process being developed to address this challenge is Adaptive Management (AM) (Holling 1978). Learning more about the resources or system to be managed and its responses to management actions, in order to develop a shift in understanding, is an inherent objective of AM (Walters, 1997; Fazey et al., 2005). Learning in AM leads to a focus on the role of feedback from the implemented actions. Such feedback-base learning models stress the need for monitoring the discrepancies between intentions and actual outcomes (Fazey et al., 2005). Monitoring becomes the primary tool for learning about the system and its performance under different management alternatives (Campbell et al., 2001).

To this aim, we assume that learning can be defined as a change in a person-system relationship, that is, the understanding of a person's place in the system and how they perceive it (Fazey et al., 2005). This definition implies that, because understanding is the goal which is achieved by the learner, each person may understand the environmental system differently and, therefore, act differently (Fazey et al., 2005). From the information production and management point of view, this implies that mental models influence an actor's perception of a problematic situation by influencing not only what data the actor perceives in the real world and what knowledge the actor derives from it (Timmerman and Langaas, 2004; Pahl-Wostl, 2007; Kolkman et al., 2005), but also what is noticed and what is taken to be significant (Checkland, 2001). It is important in information production and management that there should be a clear understanding and sharing of information users' mental models.

Therefore, contrarily to the traditional approach, in which information needs elicitation was intended in a top-down perspective, the design of a monitoring system for AM should begin by bringing together the interested parties to discuss their understanding of the system, the management problem, the information needed and how this information should be used. This implies involving a wide variety of stakeholders (i.e. scientists, managers, policy makers and members of the public at large) in a debate in which assumptions about the world are teased out, challenged, tested and discussed (Checkland, 2001), leading to the establishment of a common understanding about the system to be managed (Pahl-Wostl, 2007). This shared understanding can be structured in a system cognitive model, which allows the emergent properties of the system (i.e. variables to be monitored, thresholds, etc.) to be identified.

Among the different methods for Cognitive Modelling, an integration between Cognitive Maps (CM) and Causal Loop Diagrams (CLD) would seem particularly interesting to support monitoring system design. Given the peculiarities of the two modelling devices, CM can be used to disclose individual understanding of the system and to support the debate among participants, whereas CLD has great potentialities to simulate system dynamics.

When defining the cognitive model to be used as basis for a monitoring system, it is essential to address certain issues related to complex system dynamics. Firstly, the issue of scale must be tackled, since complex systems have structures and functions that cover a wide range of spatial and temporal scales. The impact of a given management action may vary at different scales (Campbell et al., 2001). Moreover, structures and processes are also linked across scales. Thus, the dynamics of a system at one particular scale cannot be analysed without taking into account the dynamics and cross-scale influences from the scales above and below it (Walker et al., 2006).

To deal with interaction between scales, we assume that the complex web of interacting systems can be broken down recursively into a network of individual systems, each of which determines its own fate and affects that of one or more other systems. The hierarchical structure of relationships between systems and subsystems (Campbell et al., 2001) implies that working on a particular scale often requires insights from at least two other scales, i.e. the level below, to understand the important processes that lead to the emerging characteristics of the level considered, and the level above it. Two sets of variables have to be considered for every system-subsystem pair. One set is required to describe the properties of the subsystem, whereas the second set is needed to describe the contribution of the subsystem to the performance of the whole system. This duality should be repeated at every level of the system hierarchy (Bossel, 2001).

Therefore, during the participatory process aimed at developing the cognitive model, participants should be required to think about their understanding of the total system, its essential component systems and the relationships that exist between them. The variables forming the cognitive model have to be able to describe the performance of the individual system and its contribution to the performance of the other systems. Using this inter-scale cognitive model as a basis for the design phase allows us to define a monitoring system capable of dealing with complex relationships between different scales, thus overcoming one of the main drawbacks of traditional monitoring practices.

However, adopting this inter-scale approach usually results in a demand to monitor a broader set of monitoring variables than traditional monitoring approaches. Some of these variables are fairly cheap to measure, but others, such as trends in very rare and important species, can be very expensive to monitor (Walkers, 1997). Thus, the development of an affordable monitoring program to support Adaptive Management involves substantial, scientific innovation in both method and approach, aimed at simplifying the set of monitoring variables by identifying the key components of the system.

The key components of the system, or key variables, are those that influence the system dynamics and bring about the most important changes (Walker et al., 2006; Campbell et al., 2001). Since these variables influence the overall dynamics of the system, they are of direct interest to managers, who are frequently focused on fast variables. These variables operate at different scales and with different speeds of change. The slowly changing variables determine the dynamics of the ecological system, whereas the social systems can be influenced by slow and/or fast variables (Walker et al., 2006). The conceptual models developed integrating the stakeholders' understanding of the system can be used as a basis for identifying the key variables (Campbell et al., 2001). To this aim, the analysis of CM can provided information about the relative importance of the different variables, by analysing the complexity of the causal chain. Those nodes whose immediate domain is most complex are taken to be those most central and, thus, the most important.

The identification of the key variables can also be supported by a strict integration between system monitoring and system modelling. This, in turn, is essential to any analysis of the implications of water policies. It allows the difficulties in understanding the dynamic feedback of the systems to be overcome, a particularly difficult task in an environmental context because of the number of factors involved. Moreover, humans have a limited capacity to understand the complexity of feedback in ecological systems (Fazey et al., 2005). This leads to erroneous connections between cause and effect and, thus, to erroneous conclusions about the impact of management actions. Conversely, models suggest which variables may be critical to monitor the impact of management actions, by posing elaborate hypotheses of which variables and relationships are critical to understanding the problem in question. The models then consider the dynamic implications of these hypotheses through the simulation of different scenarios. This allows monitoring networks to be designed (and re-designed) according to the model results. The potential of models to simulate future scenarios can be exploited to support the categorisation of the variables according to speed of change, i.e. slow changing variables and fast changing variables. Scenario simulation can draw attention to the role of the slow-changing variables in influencing system dynamics (Walker et al., 2006). The categorisation of variables according to speed of change can be used to program the frequency of data collection, making it easier to identify each variable's trend.

The integration between monitoring and modelling has to be considered as an iterative process. In fact, while models can simulate system dynamics, allowing the identification of key variables, the availability of new data allows the revision and updating of models. Moreover, the speed of change of the variables can also be considered iterative. Indeed, variables classified as slow changing in the model may be identified as fast changing by the monitoring system. In this case, the monitoring sample interval has to be changed. Thus, clearly a re-assessment process is needed both in models and in monitoring.

Simulation of system dynamics facilitates the identification of thresholds, which can be broadly defined as a breakpoint between two states of a system. When a threshold is exceeded, a change in system function and structure results. Such changes regard the nature and extent of feedback, resulting in changes of directions of the system itself. The changes can be reversible, irreversible or effectively irreversible (Walker et al., 2006). Two different types of thresholds can be defined, i.e. positive and negative. A positive threshold represents a desirable change in the state of the system. Such a change can be due to implemented management actions. A negative threshold can be considered as the starting point of a non-acceptable system trajectory. The recognition of these thresholds is particularly important in the case of irreversible changes. In this situation, actions are needed in order to avoid exceeding the threshold. The integration between monitoring and modelling provides information about the current state and the future trajectory of the system.

The position of the threshold is strictly linked to past experience. There are no examples where a new kind of threshold has been predicted before it has been experienced. Typically, the identification of thresholds is based on an analysis of systems similar to the one under investigation (Walker and Meyers, 2004). To this aim, a database is going to be implemented to collect empirical data on possible regime shifts in socio-ecological systems (Walker and Meyers, 2004). Some authors suggest using variances in variable trends to detect an impending system change (Brock and Carpenter, 2006). Integrating these two different approaches can be very useful. In other words, the existing experience regarding regime shifts, coming both from other systems and from the tacit knowledge of experienced and highly skilled people, can be structured and included in the system model. The variance can be calculated using monitoring data and the position of the threshold can be changed.

Integrating system modelling and monitoring iteratively highlights the importance of collecting information on trends. In fact, the availability of time series of data on the different variables allows the behaviour of the system variables and the trajectory of the system to be defined. The detection of trends can support the revision of the hypothesis concerning system dynamics, which is at the basis of the models. For these reasons it is fundamental to develop a monitoring system which is sustainable over time. To this aim, two important issues needs to be addressed, i.e. the need firstly to increase the adaptability of the monitoring system to policy and learning processes, and secondly to reduce monitoring costs through the adoption of scientific and technical innovation in information collection.

3. Adaptive monitoring and information system

Considering the issues described in the previous section, the conceptual architecture of a monitoring system for AM was defined (figure 1). From now onward, we refer to this system as Adaptive Monitoring Information System (AMIS).

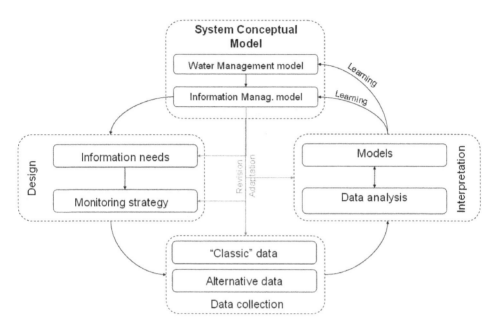

Fig. 1. AMIS conceptual architecture. The figure has been adapted from the Information cycle elaborated by Timmerman and others (2000), to emphasise the two learning processes.

As described previously, the basis for AMIS design is the conceptual model of the system, which simplifies the system and makes the key components and interactions explicit. The definition of this model is based on the integration between a participatory process, allowing experienced stakeholders to provide their understanding of the system, and models able to simulate future scenarios. The conceptual model is structured using the integration between Cognitive Maps and Causal Loop Diagrams.

Two different conceptual models, i.e. the "water management conceptual model" and the "information management conceptual model" are defined as the basis of AMIS. The former concerns the interpretation of the problem considered, while the latter concerns the information needed to solve the problem considered, and the "frames" used to interpret the information (Pahl-Wostl, 2007; Kolkman et al., 2005).

The AMIS architecture consists of four main boxes, i.e. Conceptual model elicitation, Design, Data collection and Interpretation. The links between them represent the iterative process of monitoring design, which is at the basis of AMIS. The figure was elaborated starting from the information cycle developed by Timmerman et al. (2000). This cycle depicts a framework where information users and producers communicate information needs that link the monitoring and decision processes. The monitoring program needs to be adapted to the different stages of the policy definition process, because each stage requires different types of information (Cofino, 1995; Ward, 1995) to make water management and governance adaptive.

Two possible learning processes can be identified. The first one concerns the water management conceptual model. Once information has been examined, a perspective is developed, and an insight is gained and integrated into the conceptual model itself (Kolkman et al., 2005). Information may prove initial models to be wrong and support the debate between actors, which may lead to a revision of models, through reflection and negotiation, in a social learning process. This learning may, in turn, support changes in the water management conceptual model. Moreover, feedback on management actions may generate new questions or new insights. This may make the originally agreed upon information appear inadequate, resulting in new information needs. Thus, the information needed to support a decision process evolves according to the actors' learning process, leading to revision/adaptation in monitoring strategies and data interpretation.

The second learning process relies on feedback from applied monitoring practices. As a result of experience in implementing the monitoring program and assessing its results, adaptation to monitoring may be needed (Cofino, 1995; Smit, 2003). The causes for adaptation can be found within monitoring practices: too little attention may have been spent on specifying the information needs; the information needs may have been specified in such a way that no adequate information can be produced from it, or so that it does not reflect the actual information users' needs; the selected indicators may not adequately measure what they are purported to measure; or the strategy to collect information may not have produced the right information. Furthermore, the available budgets may restrict the number of indicators that can be measured or the intensity of the network in terms of locations and frequency. New information sources may become available (e.g. progress in remote sensing technologies, etc.).

To this aim, an important innovation in AMIS concerns data collection methods. AM often results in a demand to monitor a broad set of variables, with prohibitive costs if the monitoring is done using only traditional methods of measurement. This is particularly true in developing countries, where financial and human resources are limited. In these areas, the monitoring network may cover only small part of the territory or the grid may be too sparse, making the monitoring data unsuitable for the decision process. Furthermore, traditional monitoring is costly, reducing its sustainability over time. The resulting works may be still valuable as one-off assessments, but they do not provide information about the trends of environmental resources and the evolution of environmental phenomena. Thus, the outcomes of environmental policies are often difficult to assess.

To deal with these issues, AMIS is based on the integration of alternative sources of knowledge. Thus, AMIS can be considered as the shared platform through which traditional monitoring information and innovative information sources (e.g. remote sensing monitoring, community monitoring, etc.) are integrated. Therefore, AMIS is able to adapt to data and information availability, supporting adaptive management even in data poor regions.

In Table 1, a comparison between the conventional approach and monitoring to support IWRM and AM is proposed.

Current monitoring practices	Needs for IWRM
- Based on monitoring objectives and disciplinary needs - Information users have unrealistic expectations of the information that will be produced - Data accessibility is limited - Abundant and detailed information is provided - The information provided is highly specialised - The available information is divided over various organisations - Information is transferred to the information users	- Based on policy objectives and information users' needs - The information that will be produced is jointly agreed between information users and producers - Data are publicly available and accessible - The information provided is concise and addresses the policy objectives - The information is targeted towards specific audiences - The information combines results from various organisations and is integrated over disciplines - Information is communicated to the information users and a broader stakeholder or public audience and evaluated before being incorporated into policy support
	Additional needs for AM
- The outcomes of the monitoring program (data) are the focus. - The purpose of the monitoring program is to evaluate environmental status set against target values. - Monitoring follows management and policy implementation.	- The monitoring program design and the responses on this design are as important as the results: the focus is on learning. - Monitoring becomes the primary tool for learning, i.e. understanding the system, assessing the effectiveness of management activities evaluating the system changes, and measuring progress towards participatory defined goals. - Monitoring, management and governance are interdependent.

Table 1. Comparison among current, IWRM and AM monitoring

3.1 Learning process using AMIS

Learning aspects in the AMIS are not about the monitoring as a simple process or its data, but about an increase of the system understanding, communication between stakeholders to influence decision making (McIntosh et al., 2006). While giving floor to and later using knowledge, concerns, demands, and expertise from different points of view, which result from a stakeholder involvement, one will indeed achieve better decision making with more alternatives of choice on the one hand, and a broader and more balanced acceptance of the decision making in management.

To initiate and later-on ensure learning processes using a monitoring system, all relevant stakeholder groups need access to it. Being involved when objectives are defined, data and processes transparently observed, stakeholders get enabled to learn about variables and interactions of "their own" systems and "their own" decisions which could lead to a revision or adaptation of management decisions (Pahl-Wostl, 2007. Further, this creates the feeling that stakeholders "buy in" into the product, that the monitoring system is "their" and therefore deserves more credibility (McIntosh et al., 2006). According to recent approach, the involvement of stakeholders can be extended to monitoring activities and not only to the design phase. The use of local knowledge enhances the understanding of environmental system, particularly in data poor areas. Moreover, adopting a community-based approach to monitoring can promote the public awareness of environmental issues.

Thus the intensive dialogue between science and many different stakeholders offers the opportunity for a mutual development, assessment, enhancement and implementation of new or already existing concepts, methods and tools, and helps improve the quality and acceptance of the decisions that are made. Last not least when using success-stories in management, based on the AMIS design, for the further development and enhancement of the monitoring system, the learning cycle is closed.

The following criteria, implemented into an AMIS, are indispensable to serve as a learning tool (cf. McIntosh et al., 2006):

1. **Understandability**: for each group of participants one should use "professional" indicators and perception-oriented "public" indicators to support learning processes for both of them

2. **Representativity** in involvement. Regardless of the method used to solicit user groups of the AMIS, every attempt should be made to involve a diverse group of stakeholders or broad audience that represent a variety of interests regarding the issue addressed. While key stakeholders should be invited to the process of indicator formulation, there should be also an open invitation to all interested parties to join the evaluation of the system. This adds to the public acceptance and respect of the results of the AMIS. If a process is perceived to be exclusive, both key members of the decision-making community and the wider public may reject monitoring.

3. **Scientific credibility**. Although participatory monitoring as it is understood in the AMIS design incorporates values and beliefs, the scientific components of the monitoring system must adhere to standard scientific practice and objectivity. This criterion is essential in order to maintain credibility among all groups, expert-decision-makers, scientists, stakeholders, and the public.

4. **Objectivity**. The stakeholder community must trust the facilitators of a participatory monitoring as being objective and impartial. In this regard, facilitation by university researchers or outside consultants often reduces the incorporation of stakeholder biases into the scientific components of the monitoring system.

5. **Understanding uncertainty**. Understanding scientific uncertainty is critically linked to the expectations of real world results associated with decisions made as a result of the modelling process. This issue is best communicated through direct participation in the modelling process itself.

6. AMIS' own **adaptability** to incorporate new users groups, changed frameworks and newly gained (quantitative and qualitative) data. The monitoring system developed should be relatively easy to use and up-date by the administrators. This requires excellent documentation and a good user interface. If non-scientist users cannot understand the monitoring system as a source to work with, local decision-makers will not apply it to support real management problems.

3.2 Technical adaptability of an AMIS

In this section some technical aspects related to the adaptive degree of AMIS are described.

Firstly, AMIS should be flexible and able to incorporate new information and data, of different type and with different formats. Using a relational database (RDBMS) is a sound basis to be open for new information requirements, because it is very flexible and extendable. The information can be well structured and redundancy can be avoided. The user can create new tables and link them to the existing database.

To satisfy the information needs of various user groups according to their knowledge of environmental system behaviour, different types of information for different purposes must be produced. One important aim of the AMIS is to provide the user with various methods and predefined algorithms to produce information. AMIS should provide the user with user-friendly predefined methods and algorithms to produce information, such as data visualisation tools as well as automatically generated information from incoming data.

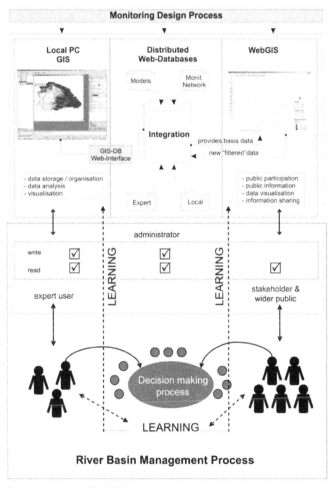

Fig. 2. Technical components of AMIS.

Another aspect of being flexible and extendable is to provide the possibility to add new modules easily, for instance hydrological or economical models, methods to analyse map layers etc. This kind of flexibility is of interest for developers or advanced users with programming skills. A modular or object oriented software structure is necessary to permit this task.

Taking the above mentioned arguments into consideration the information system is quiet flexible and open to include new information. But it is impossible to foresee what kind of requirements will be demanded from the information system in a few years. Thus, it should be possible to improve, maintain, and extend the software for everybody with programming knowledge. To be "technically sustainable" open source software should be used and local IT experts involved in the development process, particularly, if the software prototype will be produced within a project over a certain period and not by a company. One should emphasise the problem here that after a project has finished, often the developers are not available or not in charge for the product anymore. To facilitate future improvements the AMIS must be equipped with a sound documentation of the source code.

4. The adaptability of the groundwater monitoring system in Apulia Region: Main drawbacks and potential improvements

The aim of this work is to criticize the current approaches to monitoring design, highlighting the main drawbacks which hamper the adaptability of monitoring system. Moreover, potential improvements are discussed. To this aim a framework to assess the adaptability degree of monitoring design approach has been developed. The framework is structured as shown in the following table.

Criteria	Meaning
- Information producer/information users interaction	- Is the monitoring system based on the elicitation of the decision-makers' information needs?
- Degree of participation	- How many actors have been involved in the process of monitoring system design? At which level? In which phase?
- Multi-scale monitoring	- Is the monitoring system able to collect information at different spatial and temporal scale?
- Integration of information sources	- Is the monitoring system based on the integration of different sources of data and information?
- Long time sustainability	- Is the monitoring system capable to provide long time series of data?
- Monitoring/modelling interaction	- Is the monitoring system integrated with modelling to support data analysis and interpretation?
- Policy evaluation	- Is the monitoring system capable to support the evaluation of the policy impacts and suggest improvements?
- Monitoring evaluation	- Does the monitoring program provide for an evaluation and adaptation of the monitoring strategy?

Table 2. Comparison among current, IWRM and AM monitoring

This criteria have been used to evaluate the adaptability of the groundwater monitoring system of the Apulia region (Southern Italy).

The groundwater monitoring network of the Apulia Region was established in 2006 to meet the wide range of standards set by the water related national legislation adopted in 1999 (Italian Legislative Decree n. 152/1999). Consequently, the monitoring network was designed, realized and finally used in order to produce water quality and quantity information useful to characterize the environmental status of the main regional groundwater bodies.

The monitoring network has been promoted and financed by the regional offices in charge of the collection, storage and processing of data collected in accordance with relevant regulations. The network design and implementation and the enforcement of the monitoring practices fall within the scope of the project called TIZIANO whose completion is scheduled for the end of 2011.

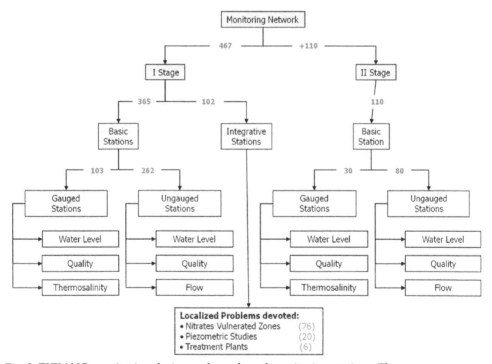

Fig. 3. TIZIANO monitoring design and number of monitoring stations. The process was composed by two main phases to identify the monitoring stations.

The TIZIANO monitoring network is made of more than 600 wells mostly spread within the boundaries of the four main aquifers of the region even if some tens of them have been located within some minor groundwater bodies. About 130 wells have been equipped with automatic probes for continuous measuring of groundwater level. During the last five years hundreds of quality and quantity measures have been made on site and thousands of samples, collected in the wells of the network, have been analyzed in laboratory in order to determine the concentration of the main chemicals, metals, organic compounds, pesticides and level of harmful microorganisms. The huge amount of information, collected during the last five years, was stored in a Geographic Information System (GIS) specifically designed for the project. It allowed regional decision-makers to assess the environmental state of the aquifers and plan and carry out specific actions to improve it, when not good, or reverse worsening trends, when they were to lead to adverse conditions of groundwater quality and quantity.

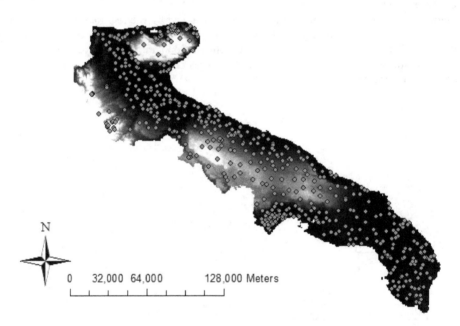

Fig. 4. Distribution of the monitoring station.

As reported above, the TIZIANO monitoring network started late in 2006, but the administrative process which led to its design and funding started several years early, at the turn of the century. In the meantime the European Union issued the Water Framework Directive (2000/60/CE), which was implemented in Italy exactly in 2006 (Italian L.D. n. 152/2006), and the, so called daughter Groundwater Directive in 2006 (2006/118/CE), recently implemented in Italy with the L.D. n. 30/2009. Although the Italian L.D. 152/1999 would herald a number of rules, then enshrined in European directives, it is evident that the future implementation of the decrees of 2006 and 2009 have clarified and modified, sometimes substantially, type, detail and timing of information to be acquired by monitoring and all management activities resulting from its processing.

4.1 Information producer/information users interaction

The Region already had a modest, monitoring network made of about 100 piezometers equipped with water level gauges, where some sporadic sampling was collected during the early 90s. Nevertheless, because of various causes, this network was abandoned after some years of functioning. At this point, within the regional offices in charge of water resources management and protection, arose the need of recovering and, possibly, potentiate the network.

In the meantime several important water related, European directives (e.g.: the Nitrate Directive, 1991/676/EEC) and national decrees had been promulgated, which forced regional water offices to move toward a detailed knowledge of the qualitative and quantitative state of water resources in order to protect such resources and restore their original natural status.

The evaluation of the institutional, legislative, technical and scientific needs and expectations led to the design of the regional groundwater monitoring network by a small team of super-experts which were careful to meet the requirements coming from various and different parts. Measures of water level and physical-chemical parameters were carried out following rules and times required by national environmental legislation implementing EU rules and a number of scientific measures and controls were preformed in order to give responses to the scientific community.

The information provided by the new monitoring system was essential, among other, in order to assess the environmental state of the Apulian groundwater bodies or delimit Nitrate Vulnerable Areas, and design and plan specific actions of different complexity and socio-economical cost, able to recover and protect groundwater.

Summarizing, measures of water level and physical-chemical parameters were carried out following rules and times required by national environmental legislation implementing EU rules and a number of scientific measures and controls were preformed in order to give responses to the scientific community.

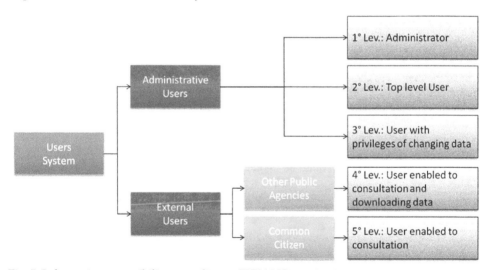

Fig. 5. Information accessibility according to TIZIANO monitoring program.

4.2 Degree of participation

From what said above derives that the position of the decision-makers in the design of the monitoring system was rather weak, i.e. the Apulian Region's role was limited to promote and fund the design. The role of decision-makers in the functioning of the TIZIANO monitoring network is strong and constant. Regional offices are in charge of producing, controlling and processing monitoring information in order to assess environmental indices and plan and execute actions for recovering deteriorated resources.

4.3 Multi-scale monitoring

Given the multi-objective frame of the monitoring each class of data has been collected with different spatial and temporal resolution. Let's have a short description of classes of data and related time-space scale starting from groundwater level.

In order to capture the cyclic behaviour of groundwater levels in the wells, measures are taken on site almost every three months. About 130 wells have been equipped with automatic water level gauges capable of acquiring and transmitting a measurement every 15 minutes. These equipped wells have been located at strategic sites, in order to use them as controlling stations. So, the project database stores groundwater levels measured at different temporal scales at different locations all over the aquifers extension. Nevertheless, there is no analysis of the inter-linkages among the process at different scales.

4.4 Integration of information sources

Given the complexity of the monitoring network, the data collection system is extremely various and includes manual and automatic measures, on site and laboratory analysis, coastal and inland exploration, airborne remote sensing. The whole amount of collected data is stored in GIS after a validation phase. Nevertheless, data coming from different platforms are sporadically integrated. The different measures follow a separated path, which passes through a separated validation step. In conclusion, the monitoring system is not based on a strong integration between sources of data.

4.5 Long time sustainability

The whole monitoring system, as currently conceived, is particularly expensive. Let's report some of the main weakness of the project concerning its own costs.

The monitoring area is objectively wide and the number of monitoring points huge, while the location of the monitoring teams is centralized and, consequently, they need to travel hundreds of kilometres during the monitoring surveys or for maintenance. Instrumentation need to be constantly maintained and often replaced due to theft.

Moreover the costs of the system are rather high due to the frequent outsourcing of monitoring activities. Costs could be reduced dramatically, if most of the monitoring practices were carried out by Regional Agencies and Offices and only very specialized activities were outsourced. In conclusion, only if an intelligent redistribution of activities within public institutions will be put in place, with a consequent cost reduction, the network is likely to become a long-term system.

4.6 Monitoring/modelling interaction

Various statistical, geostatistical, hydrogeological and hydrogeochemical, deterministic and stochastic, simple and complex models have been applied to process data collected and stored into the GIS. Nevertheless, it was not specifically designed to be compliant to any particular model. In fact, given the wide range of expected uses of the different dataset stored the design choice was to keep the organization of data extremely simple, and then easily adaptable to different kinds of models just through a simple pre-processor. In the TIZIANO monitoring project, the monitoring/modeling interaction is one-directional. That is, monitoring provides data to models, but the models are used to support the evaluation and, eventually, the re-design of the monitoring network.

4.7 Policy evaluation

Theoretically speaking, the TIZIANO groundwater monitoring system should be capable of supporting regional decision-makers at each step of the decisional path. In few words, the network should support: 1) Assessing the initial state of the natural system and reporting negative trends; 2) Controlling the effects of environmental actions and politics; 3) Alerting for undesired evolutions.

The spatial extension of the monitoring network and the number of monitoring wells should be revised at each step. Step one should be performed extensively over the monitoring area and step two should focus around risk area. Step three should be suitably designed in order to be capable of capturing any warning signal, at this step the position of the monitoring points, the parameters to be measured and the frequency of measurement need to be carefully evaluated. The TIZIANO monitoring network performed very well the first step (Assess). Unlikely, we have less evaluation elements concerning the monitoring network suitability during the second phase (Control). Finally, concerning the third step (Alert), the monitoring activity have been moved and increased around area considered mostly at risk and reduced in the rest of the region.

4.8 Monitoring evaluation

The monitoring program does not contain an evaluation phase. This means that the second learning process described in figure 1 cannot be supported.

The critical analysis of the TIZIANO groundwater monitoring system can be summarized as shown in table 3.

Criteria	Evaluation
- Information producer/information users interaction	- Mainly based on scientific requirements and legislation
- Degree of participation	- Weak in the design phase, strong during the implementation
- Multi-scale monitoring	- There is no analysis of inter-linkages between different scales
- Integration of information sources	- There is no integration
- Long time sustainability	- The monitoring costs are too high
- Monitoring/modelling interaction	- One-directional flow of information
- Policy evaluation	- The impacts on groundwater are monitored
- Monitoring evaluation	- The learning process is not supported

Table 3. Results of the evaluation

5. Conclusion

Starting from the results of the critical analysis, some drawbacks and potential improvements for the TIZIANO monitoring program have been identified and discussed in the following sections.

5.1 Main drawbacks

According to the results of the critical analysis, we can infer that the TIZIANO groundwater monitoring network cannot be considered as adaptive and it is not suitable to support the adaptive management. Firstly, the excessive cost for collecting and analyzing data have a strongly negative impact on the long term sustainability of the program. This, in turn, would reduce the capability of the monitoring system to detect the long term unintended consequences of the groundwater management policies.

Secondly, the monitoring system is not integrated in a wider program aiming to analyze the different potential impacts of the policies – e.g. socio-economic impacts. The TIZIANO monitoring program is based on the sectorial approach to environmental resources management which is still common is socio-institutional contexts characterized by a centralized and command-and-control regime. A more holistic and systemic approach is required.

Thirdly, there is no integration between different sources of information. This has a negative impact on the flexibility of the monitoring program. In fact, if the data collection is based only on traditional "static" devices – i.e. monitoring stations – then the adaptation of the monitoring program to modified information needs would be difficult: changing sensor is not always easy and/or cheap, the position of the station cannot be modified easily, even the time schedule for data collection cannot be changed easily. Although remote sensing data are mentioned in the program, the integration of this source of data with the traditional information sources is still far from being achieved.

Finally, an adaptive monitoring system requires an evaluation phase. That is, a critical analysis of the suitability of the designed monitoring system is crucial. This phase has not been considered in the current monitoring program. This means that the revision of the program depends exclusively on the political willing of the local authorities and on the availability of further funds.

5.2 Potential improvements

Some improvements to make the TIZIANO monitoring program more suitable to support the adaptive water management were defined:

- Monitoring costs: the current monitoring costs could be reduced only if an intelligent redistribution of activities within public institutions will be put in place. This means that the outsourcing activities have to be strongly reduced. Moreover, since the costs are mainly related to laboratories analysis, the integration of different sources of information would have a positive impact on monitoring costs.

- Systemic analysis of the policy impacts: the increasing awareness of the complexity of the real world forces us to adopt a system dynamic approach to monitor and analyze the different and interrelated policy impacts. Although the aim of the TIZIANO network is to collect data about the physical and chemical state of the groundwater, it has to be integrated in a more systemic monitoring program, able to detect even the socio-economical impacts.

- Integration between different sources of information: The integration of different sources of knowledge seems particularly useful to design a multi – variate and multi – scale monitoring system for adaptive management. The Use of alternative sources of information increases the flexibility of monitoring program and reduce the costs. Among the alternative sources of information, local knowledge is increasingly considered as crucial (see as example the Hyogo Framework for Action). The analysis of the literature review on this issue allowed us gain some hints. The key to guarantee the long term involvement of local community members in monitoring is to keep the monitoring activities as simple and similar to the traditional methods for environmental assessment as possible. Moreover, the involvement in monitoring is easier if the monitoring activities are incorporated in the community members' daily activities. The key to guarantee the actual usability of local knowledge in monitoring activities is: 1) fully integrating local knowledge into existing traditional institutions; and 2) structuring local knowledge so that it is transformed into meaningful and relevant information for decision-making. The integration between local and scientific knowledge allowed to enhance the reliability of local knowledge.

- Learning process in monitoring activities: as widely discussed in the scientific literature, the design of a monitoring system cannot be considered as a linear process. It is rather a cycle of design – implementation – evaluation – adaptation. The information needs can change due to several reasons. Adaptive monitoring system should be able to follow these changes. To this aim an evaluation phase should be formally included in the monitoring program. The evaluation should be based on the interaction between policy and decision makers (information users) and monitoring system managers (information producers).

6. References

Bossel, H. (2001). Assessing viability and sustainability: a systems-based approach for deriving comprehensive indicator sets. *Conservation Ecology* 5(2): 12. [online] URL: http://www.consecol.org/vol5/iss2/art12/

Brock, W. A., and S. R. Carpenter (2006). Variance as a leading indicator of regime shift in ecosystem services. *Ecology and Society* 11(2): 9. [online] URL: http://www.ecologyandsociety.org/vol11/iss2/art9/

Campbell, B., J. A. Sayer, P. Frost, S. Vermeulen, M. Ruiz Pérez, A. Cunningham, and R. Prabhu (2001). Assessing the performance of natural resource systems. *Conservation Ecology* 5(2): 22. [online] URL: http://www.consecol.org/vol5/iss2/art22/

Checkland, P. (2001). Soft System Methodology. In *Rational Analysis for a Problematic World*. Rosenhead, J., Mingers J. (eds), pp. 61-89. John Wiley and Sons, Chichester, UK.

Cofino, W.P. (1995). Quality management of monitoring programs. In *Proceeding of the international workshop on monitoring and assessment in water management; Monitoring Tailor-Made*. Adriaanse M., J van der Kraats; P.G. Stocks, and R.C. Wards (eds), 20-23 September 1994, Beekbergen, The Netherlands.

Fazey, I., J.A. Fazey, and D.M.A. Fazey (2005). Learning More Effectively from Experience. *Ecology and Society*, 10(2), 4. [online] URL: http://www.ecologyandsociety.org/vol10/iss2/art4/

Holling, C.S. (ed.) (1978). *Adaptive Environmental Assessment and Management*. John Wiley and Sons, New York.

Kolkman, M.J., M. Kok, A. van der Veen (2005). Mental model mapping as a new tool to analyse the use of information in decision-making in integrated water management. *Physics and Chemistry of the Earth*, 30: 317-332.

McIntosh, B S, Giupponi, C, Smith, C, Voinov, A, Matthews, K B, Monticino, M, Kolkman, M J, Crossman, N, van Ittersum, M, Haase, D, Haase, A, Mysiak, J, Groot, J C J, Sieber, S, Verweij, P, Quinn, N, Waeger, P, Gaber, N, Hepting, D, Scholten, H, Sulis,

A, van Delden, H, Gaddis, E, Assaf, H. (2006). Bridging the gaps between design and use: developing tools to support management and policy. In print.

Pahl-Wostl, C. (2007). The implications of complexity for integrated resources management. *Environmental Modelling and Software*, 22: 561-569.

Smit, A.M. (2003). Adaptive monitoring: an overview. *DOC Science Internal Series*, vol. 138. Department of Conservation, Wellington. 16 p.

Timmerman, J.G. and S. Langaas (2004). Conclusions. In *Environmental information in European transboundary water management*. Timmerman, J.G. and S. Langaas (eds.), pp. 240-246. IWA Publishing, London, UK. ISBN: 1 84339 038 8.

Timmerman, J.G., J.J. Ottens, and R.C. Ward (2000). The information cycle as a framework for defining information goals for water-quality monitoring. *Environmental Management* 25(3): 229-239.

Walker, B. H., L. H. Gunderson, A. P. Kinzig, C. Folke, S. R. Carpenter, and L. Schultz (2006). A handful of heuristics and some propositions for understanding resilience in social-ecological systems. *Ecology and Society* 11(1): 13. [online] URL:http://www.ecologyandsociety.org/vol11/iss1/art13/

Walker, B. and J. A. Meyers (2004). Thresholds in ecological and social–ecological systems: a developing database. *Ecology and Society* 9(2): 3. [online] URL: http://www.ecologyandsociety.org/vol9/iss2/art3

Ward, R.C. (1995). Monitoring Tailor-made: what do you want to know? In *Proceeding of the international workshop on monitoring and assessment in water management; Monitoring Tailor-Made*. Adriaanse M., J van der Kraats; P.G. Stocks, and R.C. Wards (eds), 20-23 September 1994, Beekbergen, The Netherlands.

4

ICT for Water Efficiency

Philippe Gourbesville

*Nice Sophia Antipolis University / Polytech Nice Sophia,
France*

1. Introduction

Global change poses unprecedented threats to society through impacts on both the natural environment and engineered infrastructure. Specifically, growing global population requires urban and infrastructure development at the same time as global warming demands massive investment in measures for both adaptation to future climate and mitigation through reduced emissions. The water sector is at the heart of this 21st century challenge, and the need of the hour is to have a major revision of our approaches and implementation of technology for the management of water resources, flood risk and pollution.

As mentioned recently by the Water Supply and Sanitation Technology Platform (2005) – WSSTP - representing all the European water sector actors, "water supply, storm-water drainage, wastewater collection and treatment, as well as quality and quantity management of natural water resources need to be efficiently secured or, where necessary, improved. Only through a paradigm shift from fragmented towards integrated urban water management economic development, social balance and ecological integrity can be secured. [...] During the last three decades the European water industry has built up a great competitive strength based on innovative supply and sanitation concepts, technology, knowledge and skills; availability of financial resources; wide experience in many industrial sectors; close cooperation with European R&D organisations and universities, including active involvement in R&D projects in the various European Union R&D Framework Programmes; expanding markets in the European Union and outside; European Union policy on sustainability, environment and energy; a broad spectrum of efficient governmental structures, tailored to specific local needs. The three largest companies providing water supply and sanitation services in the world are European. In addition, a large number of European Small & Medium Enterprise's (SME's) export their expertise and equipment across the world. Several European firms and institutes have prominent positions in the open market for major water and sanitation studies and implementations. The European water sector is a major economic player - 1% of GDP - with a turnover in the European Union of about 80 billion Euro and an average growth rate of 5% per year, compared to 2.5% per year average growth rate for the European Union economy."

The diagnostic provided by the profession at the European level and with the support of the WSSTP mentions that sustainable approaches for the development of water projects are needed to deliver social, economic and environmental benefits. These demands are pressing issues in the new European Member States, and in developed and developing countries outside Europe. Technologies need to be properly integrated with social, economic and organisational measures. Until now a sectoral approach in water resources management has been dominating and is still prevailing. Many actors are not fully integrated, and many stakeholders remain uninvolved. This has led to fragmented and un-coordinated implementation of policies and technologies, and often leads to inefficient or even unsustainable solutions. To achieve sustainability, Europe, as all countries, has to apply an integrated and participatory approach for water resource management. The water industry is too slow in studying and eventually adopting new technologies. The World Water Council (2009) states: "Without major technological innovations there is little hope of bringing the water equation into balance. There is no doubt that many technological changes can help improve services for millions and reduce the stress on water systems around the world."

To remain in the forefront of this competitive business, innovative skills are essential. The knowledge and experience in water supply and sanitation that is available for example in Europe is dispersed across a large number of small utilities and enterprises. Although not directly visible to the outside world, a considerable body of knowledge has been developed in designing and optimising water infrastructure and management systems over the past 150 years. This diversity of solutions adapted to local conditions in Europe is quite valuable assets in the world market. The energies of all actors in the sector must be combined to merge the dispersed knowledge and expertise and use it to enhance the competitiveness of the water sector.

The challenges faced by the water sector in Europe and worldwide are serious and well-documented. Future water shortages require immediate action on development of resources, reduction of demand and higher efficiency in treatment and transmission. Future flood risk management requires immediate action in risk assessment, defence and alleviation systems, forecasting and warning systems and institutional and governance measures. Such development requires considerable investment in research from governments and large corporations and this is now becoming apparent in many countries. The challenge is made even more difficult, however, by the requirement for solutions to be sustainable and moving towards a "low carbon economy" which are also increasingly being stipulated by government and European Union Directives. For example, the drive for higher reliability in water resource is therefore accompanied by a drive for reductions in cost, emissions, ecological and environmental impacts.

Technology has been revolutionised over recent years and now, matured with mass production allowing wider uptake of methods and devices (Gourbesville, 2009). After the development phase, technology is now entering an application and implementation phase which is targeting several fields including environment. A relevant example is given by the European Union who has defined a major priority for the next 20 years on "ICT for sustainable growth" with the ambition to lead innovation at the worldwide scale. In such context, ICT refers to technologies that provide access to information through telecommunications. It is similar to Information Technology (IT), but focuses primarily on communication technologies. This includes the Internet, wireless networks, cell phones, and other communication mediums. As defined by the European Commission, improving the quality of life should not damage the environment for future generations. Achieving sustainable growth requires better management of all natural resources, from energy to water and ICT - Information and Communication Technologies - can enable this far more efficiently (Holz, 2004), so improving environmental protection without holding back economic development.

Key concerns are the impact of climate change and the inefficient use (or over-use) of natural resources, such as drinking water and energy supplies. However, in order to achieve these objectives, the European Commission focuses its efforts on several specific areas such as Energy Efficient Buildings, Smart Electricity Grids and Smart Metering, Freight, Logistic and Transport, Greener ICT, Water Management. In this last domain, the European Union wishes to recognize the added value of ICT solutions and to support their implementation in the water domain by elaborating, validating and disseminating recommendations, guidelines and specifications on specific technologies and uses. This strategy is duplicated at the international level with the priorities of the National Science Foundation (NSF) in USA and the Green Growth project developed in South Korea.

If the diagnostic is now shared globally, it request coordinated efforts in order to implement the various ICT solution into the water sector. This sector is complex and requires a careful analysis able to underline needs and to identify the added value provided by ICT solutions according to a realistic roadmap for implementation.

2. Methodology for assessing priorities

Obviously, in the coming years the new technologies from the IT sector will affect the full water cycle and the management of the water related services. This process represents a major challenge for the 21st century. However, the impact of these new technologies – from sensors to Decision Support Systems - could be stronger and really significant if priorities are properly defined and implemented within the R&D strategies. The main driver of the strategy has to be to achieve a comprehensive architecture of an Information System (IS) dedicated to water uses and connected to others systems involved in human activities.

By definition, Information systems are implemented within an organization for the purpose of improving the effectiveness and efficiency of that organization (Silver, 1995). Capabilities of the IS and characteristics of the organization, its work systems, its people, and its development and implementation methodologies together determine the extent to which that purpose is achieved. The IS is associated to an architecture which provides a formal definition of the business processes and rules, systems structure, technical framework, and product technologies for a business or organizational information system.

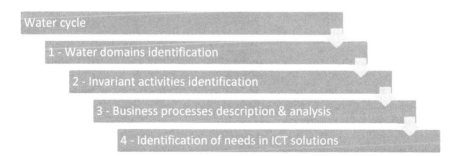

Fig. 1. General methodology for development of ICT solutions in the water sector.

In order to elaborate a specific IS for the management of the water cycle, a methodology is needed for identifying priorities and strategic investments to do in the ICT domain. The requested approach has to investigate all domains and provide a map of the various process taking places in the different domains of the water uses cycle. This formalization exercise, using mainly concepts and processes, is now requested in order to ensure the coherence of technical choices in a holistic approach.

The methodology has to start from the water cycle, to identify the various water domains and the associated activities. The activities can be then defined with business processes which can be analysed regarding the need of ICT solutions. The proposed methodology is summarized on the Figure 1.

2.1 Domains of the water cycle

The water cycle is frequently defined as the hydrologic cycle which describes the continuous movement of water on, above and below the surface of the Earth. The hydrologic cycle involves the exchange of heat energy, which leads to temperature changes and drives states of water. The water cycle figures significantly in the maintenance of life and ecosystems.

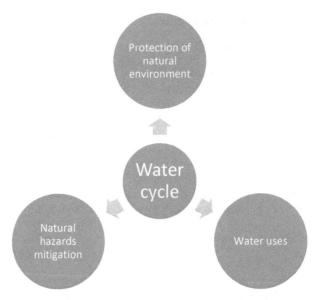

Fig. 2. Domains of water cycle.

In order to preserve this essential resource, the concept of Integrated Water Resources Management (IWRM) has been developed (Jønch-Clausen T. & Global Water Partnership (GWP), 2004). The purpose of the approach is to "promotes the coordinated development and management of water, land and related resources, in order to maximize the resultant economic and social welfare in an equitable manner without compromising the sustainability of vital ecosystems." Operationally, IWRM approaches involve applying knowledge from various disciplines as well as the insights from diverse stakeholders to devise and implement efficient, equitable and sustainable solutions to water and development problems. As such, IWRM is a comprehensive, participatory planning and implementation tool for managing and developing water resources in a way that balances social and economic needs, and that ensures the protection of ecosystems for future generations. In such approach, ICT solutions can play a key role but focus has to be given to the most demanding and relevant domains of the water cycle.

In order to identify which and how ICT solutions can be implemented, it is necessary to look at the water cycle through an approach based on functional domains and business processes. This methodology allows considering each action involved into the resource management and identifying the potential needs of ICT.

The water cycle can be divided in three domains which are associated to specific activities and business processes:

- Protection of natural environment and ecosystems;
- Natural hazards mitigation and disaster prevention;
- Water uses.

The first domain considers all actions needed to assess and advice on the environmental impacts of development proposals and projects related to specific water uses. Results are used by regulatory services. The domain covers also all conservation actions of water related ecosystems.

The second domain is focused on water related natural hazards mitigation actions. Floods, water-borne and vector disease outbreaks, droughts, landslide and avalanche events and famine are the processes covered by this domain. Every year, disasters related to meteorological, hydrological and climate hazards cause significant loss of life, and set back economic and social development by years. The disaster is defined as a serious disruption of the functioning of a community or a society causing widespread human, material, economic and/or environmental losses.

The last domain covers the added influence of human activity on the water cycle. Generally, the water uses refer to use of water by agriculture, industry, energy production and households, including in−stream uses such as fishing, recreation, transportation and waste disposal. All of those uses are directly linked to specific activities and processes which are potential targets for deployment of ICT solutions. In order to stick to the reality oft he water management operated by entities in charge of water services, the traditional classification can be reviewed. The main water uses appear then as: agriculture, aquaculture, industry, recreation, transport/navigation, and urban.

2.2 Water uses, activities and business processes

According to the defined water domains, the water uses represent the largest field where ICT solutions can be developed and implemented. The various uses may be classified and defined as follow.

- Agriculture: Irrigation water use is water artificially applied to farm, orchard, pasture, and horticultural crops, as well as water used to irrigate pastures, for frost and freeze protection, chemical application, crop cooling, harvesting, and for the leaching of salts from the crop root zone. In fact, irrigation is the largest category of water use worldwide.
- Aquaculture: Aquaculture is the farming of aquatic organisms including fish, molluscs, crustaceans and aquatic plants. Farming implies some sort of intervention in the rearing process to enhance production, such as regular stocking, feeding, protection from predators and so forth. It also implies individual or corporate ownership of the stock being cultivated. This activity uses part of the water bodies in order to develop activities.

- Industry: This water use is a valuable resource for such purposes as processing, cleaning, transportation, dilution, and cooling in manufacturing facilities. Major water-using industries include steel, chemical, paper, and petroleum refining. Industries often reuse the same water over and over for more than one purpose.
- Recreation: It often involves some degree of exercise as well as visiting areas that contain bodies of water such as parks, wildlife refuges, wilderness areas, public fishing areas, and water parks. Some of the activities that imply the uses of water for this purpose are: fishing, boating, sailing, canoeing, rafting, and swimming, as well as many other recreational activities that depend on water. Recreational usage is usually non-consumptive; however recreational irrigation such as gardening or irrigation of golf courses belongs to this category of water use. Besides, recreation and tourism represent a growing sector for industry at the worldwide scale.
- Energy: Derived from the force or energy of moving water, which may be harnessed for useful purposes, such as Energy production. There are several forms of water power currently in use or development. Some are purely mechanical but many primarily generate electricity. Broad categories include: conventional hydroelectric (hydroelectric dams), run-of-the-river hydroelectricity, pumped-storage hydro- electricity and tidal power.
- Transport/navigation: It refers to the transport of goods or people using water as a means of transportation. This water use refers only to commercial transport, since recreational transports such as sailing is considered above in Recreation water use.
- Urban: Urban water use is generally determined by population, its geographic location, and the percentage of water used in a community by residences, government, and commercial enterprises. It also includes water that cannot be accounted for because of distribution system losses, fire protection, or unauthorized uses. For the past two decades, urban per capita water use has levelled off, or has been increasing. The implementation of local water conservation programs and current housing development trends, have actually lowered per capita water use. However, gross urban water demands continue to grow because of significant population increases and the establishment of urban centres. Even with the implementation of aggressive water conservation programs, urban water demand is expected to grow in conjunction with increases in population. The urban environment is associated to a high dynamic which implies a growing complexity related to number of inhabitants and management of water resources in order to fulfil the needs of population.

The water uses are associated to business processes and are linked to economical and social values. In most of the cases, five major activities are taking place within each water use and appear as invariants. These key activities are: Investigating /surveying, observing / monitoring, designing, building and decommissioning, operating. Each activity could be defined.

- Investigating/surveying: Consists in the gathering of information of the previous and actual state and/or working of the domain in study. This assembly of information can be done either by a systematic collection of field data (survey) or a collection of information or data from a methodical research of available documents and/or the production of new ones in order to understand or to improve the actual state of the domain.

- Observing/monitoring: From a general point of view, this activity refers to be aware of the state of a system. It describes the processes and activities that need to take place to characterise and monitor the quality and/or state of the domain in study. All monitoring strategies and programmes have reasons and justifications which are often designed to establish the current status of the domain or to establish trends in its parameters. In all cases the results of monitoring will be reviewed and analysed. The design of a monitoring programme must therefore have regard to the final use of the data before monitoring starts.
- Designing (including risk assessment): Refers to the process of devising a system, component, or process to meet desired needs. It is a decision making process (often iterative) in which the basic sciences, risk assessment and engineering sciences are applied to convert resources optimally to meet a stated objective. Among the fundamental elements of the design process are the establishment of objectives and criteria, synthesis, analysis, construction, testing and evaluation. In order to obtain a design that achieves the desired needs for the domain in study, the two previous steps should have been accomplished and taken into account.
- Building & decommissioning: Consists in carrying out the proposed solution (design) for the domain. In order to execute this design, construction and/or decommission activities may be executed. It is essential a minimal environmental impact when accomplishing these activities. The tolerable environmental impact will be obtained from the risk assessment of the designing step.
- Operating: It refers to the action of manoeuvring a system. It may include the combination of all technical and corresponding administrative, managerial, and supervision actions. Operation may also include performing routine actions which keep the system in working order. This latest actions might turn out as response of problems detected during monitoring.

Fig. 3. Invariant activities taking place in the various domains and water uses.

The final step of the approach is dedicated to the identification of the various business processes which are taking place in each activity. A business process is defined as a collection of related, structured activities or tasks that produce a specific service or product (serve a particular goal) for a particular customer or customers. It implies a strong emphasis on how the work is done within an organization, in contrast to a product's focus on what. A process is thus a specific ordering of work activities across time and place, with a beginning, an end, and clearly defined inputs and outputs: a structure for action. Some processes result in a product or service that is received by an organization's external customer. These are called primary processes. Other processes produce products that are invisible to the external customer but essential to the effective management of the business. These ones are called support processes. In keywords, a business process has a goal, has specific inputs and specific outputs, uses resources, has a number of activities that are performed in some order, may affect more than one organizational unit - horizontal organizational impact - and creates value of some kind for the customer. An example of a business process for a water utility can be meter reading. It has to be done in concordance of the billing period. The goal of this process is to give inputs to the billing department, and see the progress of the customer's consumption. Depending on the technology used for the metering (smart or manual metering), different resources (technology, personnel) are used.

The uses in urban environment, carried out by water utilities, can be defined with a limited number of business processes – 29 in total - summarized into the table 1 and which are covering drinking water, waste water and storm water management. The final step of the approach is then to identify for each business process how ICT solutions can be implemented and provide added value. This diagnostic has to be shared by professionals and operators in order to ensure a coherent deployment. This validation process can be made through an associative body gathering representatives from all involved sectors.

1 - Asset management	16 - Water primary network management and water balance
2 - Crisis management	17 - Water secondary network management
3 - Field intervention management	18 - Leak detection
4 - Field works	19 - Meter reading (AMR & MMR)
5 - Use of GIS	20 - AMR & MMR management
6 - Maintenance of GIS	21 - Public service contract management
7 - Management of plant maintenance	22 - Waste water network management
8 - Electro mechanical maintenance	23 - Storm water network management
9 - Laboratory activity and quality control	24 - Waste water treatment plant management
10 - Automation & sensors	25 - Sewer inspection and sewer cleaning
11 - Real time network management	26 - Billing
12 - Planning and design of new assets and plants	27 - Customer care & communication
13 - Water resources management	28 - Innovation & pilots
14 - Environment management	29 - Supports
15 - Drinking water treatment plant management	

Table 1. Business processes for urban uses.

3. The @qua approach

The European Union has defined a key objective for his industrial development on interoperability of systems. This approach is dedicated to various domain including environment and water. In order to support this vision, the European Commission has launched a Thematic Network called @qua under the CIP-ICT PSP Programme. The ICT Policy Support Programme (ICT PSP) under the Competitiveness and Innovation Programme (CIP) aims at stimulating innovation and competitiveness through the wider uptake and best use of ICT by citizens, governments and businesses, particularly Small and Medium-sized Enterprises (SMEs). The approach is based on leveraging innovation in response to growing societal demands.

In his programme frame of ICT Policy Support Programme (ICT PSP) 2011, the General Direction Information Society (DG INFSO) of the European Commission has launched a new theme network dedicated to Innovation Communication Technologies for water management. This domain represents a sector which the European Union wishes to develop during the next 10 years and it's contemplated in different initiatives of the Digital Agenda for Europe 2020 which will allow at the same time improving the user's services quality and developing a sustainable management of resources. These objectives will be achieved with the improvement of already available technologies, adaptation of the existing solutions and the identification of R&D axes to work on the next years.

@qua Innovation Network (http://www.a-qua.eu), founded by 17 partners and managed by Nice Sophia Antipolis University gathers thus ICT and water services leading actors from SME to majors, research entities developing competences in both sectors, local and regional authorities directly responsible for water policy and water management. Partners have developed significant expertise about the interface of ICT and water and at the same time, covering the full spectrum of the water related domain. @qua provides a forum to exchange and to share expertise in deploying innovative ICT solutions for water management, studies feasibility of standardized ICT solutions and interoperability in the field of water management across the EU and develops specifications and guidelines according to a jointly defined "level of sharing" among representatives of professional sectors. Focus of @qua is on gathering and sharing experiences on how to overcome barriers to the introduction of ICT solutions for innovative water management and on how to ensure their wider uptake and best use. Partners have the ambition to develop and to promote the interoperability principle and the use of common standards in the water industry. In a holistic and consistent approach, @qua addresses all the issues of the water management from resources to societal changes, using a wide range of ICT solutions: data acquisition, numerical modelling, real-time monitoring and field operation management.

3.1 The @qua methodology

The @qua thematic network members have developed a general methodology based around few steps which can be summarized as follow:

Step 1.   Water business processes and ICT solutions: identification of gaps and expectations of the water domain professionals on ICT solutions;

Step 2.   Identification and validation of innovative ICT solutions by the ICT professionals with the objective to bridge the identified gaps during the Step 1;

Step 3.   Develop the "level of sharing" of each ICT solution in order to address interoperability, standards, architecture and roadmap for implementation issues;

Step 4.   Produce guidelines, standards and specifications on specific ICT solutions needed
          by the water domain in order to achieve a more efficient water management.
The two main characteristics of the defined approach are:
- the global analysis based on "business processes" and associated added value;
- the definition and the use of concept of "level of sharing" to decide which ICT
  innovations could be widely disseminated throughout the water profession.

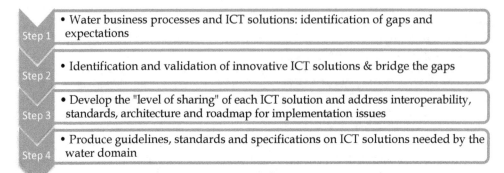

Step 1 • Water business processes and ICT solutions: identification of gaps and expectations

Step 2 • Identification and validation of innovative ICT solutions & bridge the gaps

Step 3 • Develop the "level of sharing" of each ICT solution and address interoperability, standards, architecture and roadmap for implementation issues

Step 4 • Produce guidelines, standards and specifications on ICT solutions needed by the water domain

Fig. 4. The @qua methodology.

The initial step, led by water utilities and water engineering companies, is dedicated to the analysis of the business processes, both for the artificial cycle and the natural cycle of water, and both for design and for operations. The business processes are described at a macro scale, where the tiny differences between entities are not seen and where just the common "backbone" is visible. These business models are used as "base maps" in order to show the unequipped - or poorly equipped - steps in terms of ICT. A special attention is turned to the analysis of added value of these unequipped steps. The diagnostic characterizes the added value not only on the economic point of view, but also on sociological and ecological dimensions. In addition to the common map of the water business processes itself, the result of this step is the list of the steps / processes that "deserve" to be equipped with new ICT tools. This effort of analysis according to the business processes vision represents an essential input in the water domain. Until now this diagnostic was not established for several reasons and especially due to the low maturity of water industrial domain regarding ICT solutions and uses.

The second step is led by the ICT sector representatives and consists in a technologic analysis of the needs and requests written by the water companies' representatives. The step includes not only the assessment of the feasibility, the potential availability and the cost of the requests, but it will also propose other tracks, unimagined or not foreseen during the previous step. The water companies have a partial vision of ICT solutions and they need a better knowledge of the current trends of the ICT industry / market. Alternating the leadership of the steps between the "water people" - water companies and other stakeholders - and the "ICT people" brings an efficient synergy.

The third step is focused on the determination of the "level of sharing". This concept is a central element which is developed and used by the @qua network. For the time being, the use and the implementation of existing ICT solutions in the water domain is made case by case, with a quite variable customization which is covering a simple technical adaptation like wavelength, to in depth R&D development like the use of alternative energy sources for power supply in waste water monitoring actions. The partners of the @qua network have significant experience of implementation and development actions. The spectrum of their expertise is covering most the business processes involved in the water domain. From this experience and according to their identified needs in innovative ICT solutions, they define, for each technology identified as a priority, the requested level for developing an efficient interface between the different components involved into the business process. Such work represents a major output for the @qua network and constitutes clearly an added value provision by the network to various professional communities. It is clear that in a wide community as the European water profession, the status of the various Information Systems has a very high variety. This step will analyse the "IS/IT context" parameters in the profession: maturity of the IS, level of integration (integration of the IS itself as well as integration in the business processes), level of alignment with the strategy, and the local parameters (ERP/ software already installed, other relevant IT projects, trends of the local IS/IT market, etc.). This step proposes the ideal "level of sharing", i.e. the level which will maximize the effectiveness and efficiency of the new ICT tools by taking into account the actual current IT/IS situation. This output defines the outcomes of the @qua network, which could go from the very theoretical - methodologies, data models, architectures, principles of standardization, etc. - to the very concrete elements such as list of devices compliant with the selected telecom standards, deployment of a common software and instructions of customization, etc.

In a final step, the production of the guidelines and specifications whose needs are identified in the previous steps. According to the results of the previous step, these results can go from very generic guidelines to more precise technical specifications such as hardware requirement for sensors, software architecture, strategy for implementation and deployment in water services, metadata architecture, business process description and standards. A similar approach has been partly applied with HarmonIT project (http://www.harmonit.org) on the specific field of the hydroinformatic systems interoperability and the development of the OpenMI standards (http://www.openmi.org). In the case of the @qua approach, the spectrum is much more wider because it's addressing most of the business processes involved in all water uses and domains.

3.2 The expected results and impacts

The water domain - and water stakeholders - is very wide and covers a huge number of business processes especially if all domains and activities are considered. This situation legitimates the mapping process and the prioritization of gaps that need to be bridged. Clearly the efforts have to be focused on five major areas directly linked to the urban water use which where both expectations and possibilities are the highest:

a.   Real time monitoring
   - Specially real time networks monitoring including Automated Meter Reading(AMR);
   - Installation of leak detectors in the network;
   - Real time quality management (disinfectant, turbidity, pH, temperature, conductivity, RedOx, etc.);
   - Sensors at all Points Of Use (POU);
   - Real time information of customers and stakeholders;
   - Related technologies such as Supervisory Control And Data Acquisition (SCADA), GIS, telecommunications, sensors (especially low cost sensors), inverse models, decision support systems.

b. Cities of Tomorrow
    - In the current vision , there is an absolute need of generalized ICT in the operation of the cities of the future, or sustainable cities, or water-sensitive cities;
    - Cascading usages of water (incl. re-use and recycling), rainwater harvesting, storm water management, desalination, managed aquifer recharge, micro treatment plants, etc. are the core techniques of the cities of the future These techniques need a very high level of monitoring and thus, a sophisticated density of ICT;
    - Leakage reduction in distribution networks;
    - Improving water efficiency in cities.
c. Asset Management and Field Work Management
    - In-pipe and "through road" condition assessment sensing technologies;
    - Continuous performance, condition and risk assessment sensors and prediction models;
    - Optimised network operation and "just in time" repairs and investment programmes;
    - GIS/GPS information;
    - Buried asset electronic identification and tagging devices, wireless communication through road materials;
    - "Wearable computers" for field workers, giving access in real time to all data bases of the company, with interfaces consistent with field conditions.
d. Energy Efficiency
    - Smart grid in water distribution systems (real time management of pumping strategy, refined demand forecast, optimization of network management and of operating costs);
    - Tools for energy saving in treatment plants;
    - Real time status monitoring (open/closed) of manual valves (cf. above : equipment of field operators);
    - Monitoring and control of heat recovery in wastewater;
    - Tools for Smart Metering / Smart Pricing (e.g. condition-based tariffs).
e. Water efficiency
    - Improving water efficiency in cities;
    - Improving water efficiency in agriculture, including detection of illegal abstraction;
    - Ecosystems and land-use management in perspective of project scope and available resources.

4. Some ICT solutions for water efficiency

The analysis of the domains and the business processes demonstrates the relevance and the key position of the data acquisition process through sensors located in the various sectors of the water cycle. This need is recurrent and could be seen in the three domains and takes a central position in surveying, monitoring and operating activities.

4.1 The sensor revolution

The analysis of the domains and the business processes demonstrates the relevance and the Following the PC revolution in the 1980s and the Internet revolution in the 1990s, the on-going revolution is connecting the Internet back to the physical world, creating that world its first electronic nervous system or Information System. The sensor revolution is based on devices that monitor environment - natural & built - in ways that could barely imagine a few years ago.

A sensor is any device that can take a stimulus, such as heat, light, magnetism, or exposure to a particular chemical, and convert it to a signal. Sensors have certainly been around for a very long time with scales (weight sensors), thermometers (temperature sensors) and barometers(pressure sensors). More recently, scientists and engineers have come up with devices to sense light (photocells), sound (microphones), ground vibrations (seismometers), and force (accelerometers), as well as sensors for magnetic and electric fields, radiation, strain, acidity, and many other phenomena.

While the concept of sensors is nothing new, the technology of sensors is undergoing a rapid transformation. Indeed, the forces that have already revolutionized the computer, electronics, and biotech industries are converging on the world of sensors from at least three different directions:

Smaller. Rapid advances in fields such as nanotechnology and (micro electro-mechanical systems (MEMS)) have not only led to ultra-compact versions of traditional sensors, but have inspired the creation of sensors based on entirely new principles. The reduced size fits perfectly with the constraints of the water supply and open possibilities into the monitoring and operating activities.

Smarter. The exponentially increasing power of microelectronics has made it possible to create sensors with built-in "intelligence." In principle, at least, sensors today can store and process data on the spot, selecting only the most relevant and critical items to report. One of the emerging concepts in this domain is the ubiquitous computing paradigm. This approach is highly relevant for the water domain especially for all warning and monitoring systems which may avoid the centralized design.

More Mobile. The rapid proliferation of wireless networking technologies has cut the tether. Today, many sensors send back their data from remote locations, or even while they're in motion.

In the urban water domain, the new sensors are already deeply impacting several business processes with Automated Meter Readers (AMR), water quality control devices and operating supervision. Such trend is following the recent evolution observed in energy distribution sector. An emblematic evolution is the one taking place with the introduction of the smart metering concept for water consumption monitoring.

4.2 From mechanical meters to smart metering

Water meters reading remains one of the core business process of water utilities or public services in charge of drinking water supply. This activity requests a good level of organization and a good management of the devices. To date, water meters have been accumulation meters, pulse meters or interval meters which are all mechanical devices. The data are collected directly regularly on the field. This process can report about consumption and can detect some leakages into the network. However, reactivity is low due to the limited visits on the field. The past decade has seen an evolution of conceptual design of advanced or smart metering and its terminology. Driven by electricity investment, metering has evolved from accumulation meters to interval meters with simple communications, to advanced or smart metering with an increased range of metering functionality. This increase in electricity meter functionality and complexity has started to be mirrored in the water industry.

Interval metering is comparatively more expensive than pulse metering, as the interval meter is required to constantly monitor the water flows through the meter and record this volume at the expiration of the metering interval. By using a fine pulse quantum and analysing the time stamps of these pulses, pulse metering data can be used to approximate interval water metering data and hence deliver similar benefits. Use of pulse metering where a time stamp is made when a certain quantum of water is consumed, is more common in the water industry and these pulse meters are available at reasonable cost.

Smart water metering for the water industry will extend beyond the capability of Automated Meter Reading (AMR). Smart water metering is expected to, as a minimum, establish more granular - within a day - water usage data, two-way communications between the water utility and the water meter, and potentially include communications to the customer. With respect to a customer's household, smart water metering could enable:

- Recording of water consumption within a day;
- Remote meter reading on a scheduled and on-demand basis;
- Notification of abnormal usage to the customer and/or the water utility;
- Control of water consumption devices within a customer's premise;
- Messaging to the customer;
- Customised targeting of segments.

The options to be considered for smart water metering are:

- Choice of communication to the water authority/water utility and the home;
- Choice of consumption data measurement (pulse or interval metering).

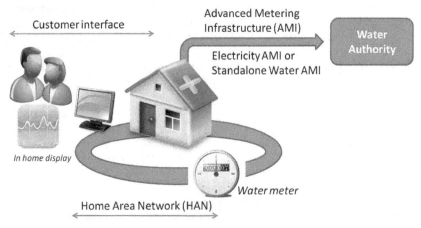

Fig. 5. Smart water metering logical architecture.

Options for the implementation of smart water metering communications arise through choices on:

- Water authority/water utility communications: The method and frequency of data collection through either drive-by collection, leveraging electricity Advanced Metering Infrastructure (AMI) communication networks or standalone water AMI communications networks;
- Customer communications: The method of communicating consumption information to customers: either in real-time across a Home-Area-Network (HAN), or in a historical manner through bills.

Since 2006, various pilot projects - from 100 to 500 smart meters – have been implement worlwide and espacially in Europe within France, Italy, Spain and Malta. The projects are carried out by the water utilities who are supporting development and implementation in various municipalities and for different situations (type of building, type of cities, ...). Most of the projects are based on wireless devices and very few are deployed on the wire networks. Following the first experiments, the main water utilities have already initiated the implementation of smart meters at a large scale with for example more than 350 000 units for France.

The pilot studies and experiments carried out since several years by the water utilities have demonstrated the savings in water consumption due to the use of the smart metering. The savings are taking place at various levels such as:

- Reduction of individual consumption. The details of the consumption are accessible through various media such as a specific website or a small electronic terminal. The information provided to the consumer immediately generates a reduction up to 15%;
- Reduction of water consumption at the macro scale (city to block). The smart metering allows to identified non conform water consumption and consequently help to reduce leakages after and before the smart meter itself. Text messages could be sent to consumers when the consumption is initiating a non coherent pattern with the previous consumption. The water utilities can also detect major leakages on the networks.
- The knowledge in real time of the water consumption allows to identify seasonal needs of the population and to anticipate the volumes of resources to mobilize. This approach allows a more functional use of resources and contributes globally to reduce the consumption.
- The knowledge in real time of the water consumption opens the doors to a new approach about pricing, based on seasonal and even hourly values.

Today, according to various publications and sources (Oracle, 2011), about a third of water utility managers in USA say they are in the early stages of adopting smart meters, despite the fact that 71 percent of water users say that having more detailed information on their water consumption would promote better water conservation. This figure is representative of the worldwide situation. From the water utilities point of view, the following benefits to adopting smart meters could be identified:

- enabling early leak detection ;
- supplying customers with tools to monitor/reduce water use;
- providing more accurate water rates;
- curbing overall water demand;
- improving the ability to conduct preventative maintenance.

The financial efficiency of the smart metering has been already demonstrated through various study cases and pilots (Marshment Hill Consulting, 2010) In developing countries where development of infrastructures and management of water resources represent a great challenge, the opportunity to invest in the smart metering concept is clearly a key issue which request an integrated effort in the global urban management.

5. Conclusion

The water sector represents a major challenge for the 21st century. The climate evolution combined with the growing of pressure of populations will generate new stresses on a limited resource which has to be carefully managed and protected. The fast development of ICT solutions allows today to enter a new area which may be characterized by the idea to move from a scarcity of data to a continuous flow of data - "data rich world" - about natural and built environment. This new situation will become a reality in the coming two decades and will allow potentially improving, globally, the water management. However, if this perspective represents a clear benefit both for natural and manmade environments, it request the development of a coherent vision based on a process allowing to integrate the fragmented activities developed until now in the water sector. The ICT solutions will allow this integration process but they have to be coordinated under guidelines and standards which have to be jointly defined by the various actors of the water sectors. Regulating bodies, public services, water utilities and IT producers are invited through organisations like @qua, to engage an active dialog in order to develop a coherent strategy. The suggested approach, based on business processes, represents a solution which has to be extended to all activities and domains of the water sector. It implies a real mobilization of all actors from who have to formalize their processes. Of course this effort requests a maturity in the process itself in order to be able to characterize the tasks and their dynamic.

The water sector represents a vast area where ICT solutions can be implemented and provide a real improvement. In order to benefit of these solutions, the water sector has to be pro active and structured in order to express needs. This challenging and exciting task will mobilize many professionals from both sectors and will request debates within the society on choices regarding water and its management.

6. Acknowledgment

The @qua thematic network and this work is funded under the ICT Policy Support Programme of the 7th Framework Program (FP7) of the European Commission.

7. References

Gourbesville, P. (2009) *Data & hydroinformatics: new possibilities and new challenges.* Journal of Hydroinformatics, Vol 11 No 3–4 pp 330–343, ISSN: 1464-7141

Jønch-Clausen T. & Global Water Partnership (GWP) (2004) *IWRM and Water Efficiency Plans by 2005: Why, What and How?*, GWP, 45p, Sweden, ISSN: 1403-5324

Holz, K.P., Hildebrandt G., Weber L. (2006) *Concept for a Web,-based Information System for Flood Management*, Natural Hazards, 38, pp 121–140, ISSN: 0921-030X

Marshment Hill Consulting (2010) *Smart Water Metering Cost Benefit Study*, Marshment Hill Consulting, Melbourne, Available from: http://www.water.vic.gov.au/__data/assets/pdf_file/0003/61545/smart-water-metering-cost-benefit-study.pdf

Oracle (2011) Smart Grid Challenges & Choices, Part 2: North American Utility Executives' Vision and Priorities, Oracle, USA, Available from: http://www.oracle.com/us/dm/h2fy11/utilities-survey-report-400044.pdf

Silver, M.S.; Markus M. L.; Mathis Beath C. (1995) The Information Technology Interaction Model: A Foundation for the MBA Core Course, *MIS Quarterly*, Vol. 19, No. 3, Special Issue on IS Curricula and Pedagogy (Sep., 1995), pp. 361-390, ISSN 1937-4771

Water Supply and Sanitation Technology Platform – WSSTP (2005) *Water Safe, strong and sustainable. European vision on water supply and sanitation in 2030*, WSSTP, Brussels, ISSN-1725-390X

World Water Council (2009) *Politics gets into water. Triennal report 2006-2009*, World Water Council, Marseille. *Available from:* http://www.worldwatercouncil.org/fileadmin/wwc/Library/Publications_and_reports/Activity_reports/TriennalReport_2006-2009.pdf

5

Autonomous Decentralized Control Scheme for Long-Term Operation of Large Scale and Dense Wireless Sensor Networks with Multiple Sinks

Akihide Utani
Tokyo City University,
Japan

1. Introduction

Various communication services have been provided. They include environmental monitoring and/or control, ad-hoc communication between mobile nodes, and inter-vehicle communication in intelligent transport systems. As a means of facilitating the above advanced communication services, autonomous decentralized networks, such as wireless sensor networks (Akyildiz et al., 2002; Rajagopalan & Varshney, 2006), mobile ad-hoc networks (Perkins & Royer, 1999; Johnson et al., 2003; Clausen & Jaquet, 2003; Ogier et al., 2003), and wireless mesh networks (Yamamoto et al., 2009), have been intensively researched with great interests. Especially, a wireless sensor network, which is a key network to construct ubiquitous information environments, has great potential as a means of realizing a wide range of applications, such as natural environmental monitoring, environmental control in residential spaces or plants, object tracking, and precision agriculture (Akyildiz et al., 2002). Recently, there is growing expectation for a new network service by a wireless sensor network consisting of a lot of static sensor nodes arranged in a service area and a few mobile robots as a result of the strong desire for the development of advanced systems that can flexibly function in dynamically changing environments (Matsumoto et al., 2009).

In this chapter, a large scale and dense wireless sensor network made up of many static sensor nodes with global positioning system, which is a representative network to actualize the above-mentioned sensor applications, is assumed. In a large scale and dense wireless sensor network, generally, hundreds or thousands of static sensor nodes limited resources, which are compact and inexpensive, are placed in a service area, and sensing data of each node is gathered to a sink node by inter-node wireless multi-hop communication. Each sensor node consists of a sensing function to measure the status (temperature, humidity, motion, etc.) of an observation point or object, a limited function of information processing, and a simplified wireless communication function, and it generally operates on a resource with a limited power-supply capacity such as a battery. Therefore, a data gathering scheme and/or a routing protocol capable of meeting the following requirements is mainly needed to prolong the lifetime of a large scale and dense wireless sensor network composed of hundreds or thousands of static sensor nodes limited resources.

1. Efficiency of data gathering
2. Balance of communication load among sensor nodes
3. Adaptability to network topology changes

As data gathering schemes for the long-term operation of a wireless sensor network, cluster-ing-based data gathering (Heinzelman et al., 2000; Dasgupta et al., 2003; Jin et al., 2008) and synchronization-based data gathering (Wakamiya & Murata, 2005; Nakano et al., 2009; Nak-ano et al., 2011) are under study, but not all the above requirements are satisfied. Recently, bio-inspired routing algorithms, such as ant-based routing algorithms, have attracted a sign-ificant amount of interest from many researchers as examples that satisfy the three require-ments above. In ant-based routing algorithms (Subramanian et al., 1998; Ohtaki et al., 2006), the routing table of each sensor node is generated and updated by applying the process in which ants build routes between their nest and food using chemical substances (pheromon-es). Advanced ant-based routing algorithm (Utani et al., 2008) is an efficient route learning algorithm which shares route information between control messages. In contrast to conven-tional ant-based routing algorithms, this can suppress the communication load of each sen-sor node and adapt itself to network topology changes. However, this does not positively ease the communication load concentration on specific sensor nodes, which is the source of problems in the long-term operation of a wireless sensor network. Gradient-based routing protocol (Xia et al., 2004) actualizes load-balancing data gathering. However, this cannot su-ppress the communication load concentration to sensor nodes around the set sink node. Int-ensive data transmission to specific sensor nodes results in concentrated energy consumpti-on by them, and causes them to break away from the network early. This makes long-term observation by a wireless sensor network difficult.

In a large scale and dense wireless sensor network, the communication load is generally co-ncentrated on sensor nodes around the set sink node during the operation process. In cases where sensor nodes are not placed evenly in a large scale observation area, the communica-tion load is concentrated on sensor nodes placed in an area of low node density. To solve this communication load concentration problem, a data gathering scheme for a wireless sen-sor network with multiple sinks has been proposed (Dubois-Ferriere et al., 2004; Oyman & Ersoy, 2004). In this scheme, each sensor node sends sensing data to the nearest sink node. In comparison with the case of one-sink wireless sensor networks, the communication load of sensor nodes around a sink node is reduced. In each sensor node, however, the destinati-on sink node cannot be selected autonomously and adaptively. In cases where original data transmission rate from each sensor node is not even, therefore, the load of load-concentrated nodes is not sufficiently balanced. An autonomous load-balancing data transmission scheme is required.

This chapter represents a new data gathering scheme with transmission power control that adaptively reduces the load of load-concentrated nodes and facilitates the long-term operati-on of a large scale and dense wireless sensor network with multiple sinks (Matsumoto et al., 2010). This scheme has autonomous load-balancing data transmission devised by consider-ing the application environment of a wireless sensor network as a typical example of compl-ex systems where the adaptive adjustment of the entire system is realized from the local int-eractions of components of the system. In this scheme, the load of each sensor node is auton-omously balanced. This chapter consists of four sections. In Section 2, the above data gather-ing scheme (Matsumoto et al., 2010) is detailed and its novelty and superiority are described. In Section 3, the results of simulation experiments are reported and the effectiveness of our scheme (Matsumoto et al., 2010) is demonstrated by comparing its performances with those of existing schemes. In Section 4, the overall conclusions of this work are given and future problems are discussed.

2. Autonomous decentralized control scheme

To facilitate the long-term operation of an actual sensor network service, a recent approach has been to introduce multiple sinks in a wireless sensor network (Dubois-Ferriere et al., 20-04; Oyman & Ersoy, 2004). In a wireless sensor network with multiple sinks, sensing data of each node is generally allowed to gather at any of the available sinks. Our scheme (Matsumoto et al., 2010) is a new data gathering scheme based on this assumption, which can be expected to produce a remarkable effect in a large scale and dense wireless sensor network with multiple sinks. In our scheme, each sensor node can select either of high power and low power for packet transmission, where high power corresponds to normal transmission power and low power is newly introduced to moreover balance the load of each sensor node.

2.1 Routing algorithm

Each sink node has a connective value named a "value to self", which is not updated by transmitting a control packet and receiving data packets. In the initial state of a large scale and dense wireless sensor network with multiple sinks, each sink node broadcasts a control packet containing its own location information, ID, hop counts(=0), and "value to self" by high power. This control packet is rebroadcast throughout the network with hop counts updated by high power. By receiving the control packet from each sink node, each sensor node can grasp the "value to self" of each sink node, their location information, IDs, and the hop counts from each sink node of its own neighborhood nodes.

Initial connective value of each sensor node, which is the connective value before starting data transmission, is generated by using the "value to self" of each sink node and the hop counts from each sink node. The procedure for computing initial connective value of a node (i) is as follows:

1. The value [$v_{ij}(0)$] on each sink node (j=1, ... ,S) of node (i) is first computed according to the following equation

$$v_{ij}(0) = vo_j \times dr^{hops_{ij}} \quad (j = 1,...,S)$$ (1)

where $vo_j (j$=1, ... ,S) is the "value to self" of sink node (j), $hops_{ij}(j$=1, ... ,S) is the hop counts from sink node (j) of node (i). dr represents the value attenuation factor accompanying the hop determined within the interval [0,1].

2. Then, initial connective value [$v_i(0)$] of node (i) is generated by the following equation

$$v_i(0) = \max v_{ij}(0) \quad (j = 1, ... ,S)$$ (2)

where this connective value [$v_i(0)$] can be also conducted from the following equation

$$v_i(0) = vm_i(0) \times dr$$ (3)

In the above Equation (3), $vm_i(0)$ represents the greatest connective value before starting data transmission in neighborhood nodes of node (i).

Before data transmission is started, each sensor node computes initial connective value of each neighborhood node based on the above Equations (1) and (2), and stores the computed connective value, which is used as the only index to evaluate the relay destination value of each neighborhood node, in each neighborhood node field of its own routing table.

2.2 Data transmission and connective value update

For a while from starting data transmission, each sensor node selects the neighboring node with the greatest connective value from its own routing table as a relay node, and transmits the data packet to this selected node by high power. In cases where more than one node shares the greatest connective value, however, the relay node is determined between them at random. The data packet in each sensor node is not sent to a specified sink node. By repetitive data transmission to the neighboring node with the greatest connective value, data gathering at any of the available sinks is completed. In our scheme, the connective value of each sensor node is updated by considering residual node energy. Therefore, by repetitive data transmission to the neighboring node with the greatest connective value, the data transmission routes are not fixed.

To realize autonomous load-balancing data transmission, in our scheme (Matsumoto et al., 2010), the data packet from each sensor node includes its own updated connective value. We assume that a node (l) receives a data packet at time (t). Before node (l) relays the data packet, it replaces the value in the connective value field of the data packet by its own renewal connective value computed according to the following connective value update equation

$$v_l(t) = vm_l(t) \times dr \times {e_l(t)}/{E_l} \tag{4}$$

where $vm_l(t)$ is the greatest connective value at time (t) in the routing table of node (l). $e_l(t)$ and E_l represent the residual energy at time (t) of node (l) and the battery capacity of node (l), respectively.

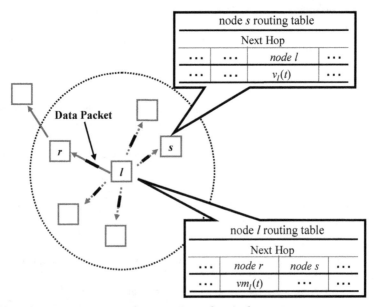

Fig. 1. Data packet transmission and connective value update

In our scheme, the data packet addressed to the neighboring node with the greatest connective value is intercepted by all neighboring nodes. This data packet includes the updated connective value of the source node based on the above Equation (4). Each neighborhood node that intercepts this packet stores the updated connective value in the source node field of its own routing table. Fig.1 shows an example of data packet transmission and its accompanying connective value update. In this example, node (l) refers to its own routing table and addresses the data packet to node (r), which has the greatest connective value [$vm_l(t)$]. When this data packet is intercepted, each neighboring node around node (l) stores the updated connective value [$v_l(t)$] in the data packet in the node (l) field of its own routing table.

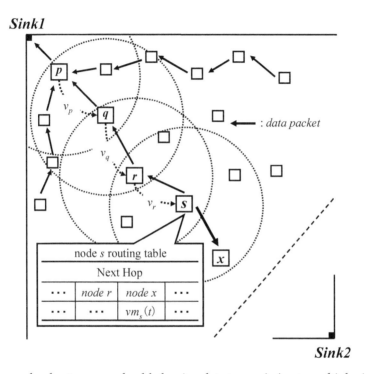

Fig. 2. An example of autonomous load-balancing data transmission to multiple sinks

Our scheme (Matsumoto et al., 2010) requires the construction of a data gathering environment in the initial state of a large scale and dense wireless sensor network with multiple sinks, but needs no special communication for network control. The above-mentioned simple mechanism alone achieves autonomously adaptive load-balancing data transmission to multiple sinks, as in Fig.2. The lifetime of a wireless sensor network can be extended by reducing the communication load for network control and solving the node load concentration problem.

2.3 Transmission power control

For data packet transmission, the transmission power of each sensor node is switched to low power if its own residual energy is less than the set threshold [T_e]. In this case, each sensor node selects the neighboring node with the greatest connective value within range of radio wave of low power as a relay node, and transmits the data packet to this selected node by low power.

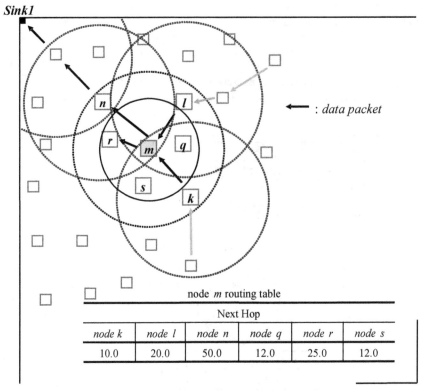

Fig. 3. An example of transmission power control

Fig.3 shows an example of the above transmission power control, which means that the tra-nsmission power of each sensor node is switched to low power according to the above con-dition. In this example, node (m) is a load concentration node. Node (m) has autonomously transmitted the data packet to node (r) with the greatest connective value within low power range by low power because its own residual energy has become less than the set threshold [T_e]. By switching to low power, the energy consumption of node (m) is saved, but node (k) and node (l) may continue to transmit the data packet to node (m) because they cannot grasp the updated connective value of node (m). In our scheme, therefore, every tenth data packet from the node switched to low power is transmitted by high power.

3. Simulation experiment

Through simulation experiments on a wireless sensor network with multiple sinks, the perf-ormances of our scheme have been investigated in detail to verify its effectiveness.

3.1 Conditions of simulation
In a large scale and dense wireless sensor network with multiple sinks consisting of many static sensor nodes placed in a large scale observation area, only sensor nodes that

detected abnormal data set were assumed to transmit the measurement data. The
conditions of the si-mulation which were used in the experiments performed are shown in
Table1. In the initial state of the simulation experiments, static sensor nodes are randomly
arranged in the set ex-perimental area, and multiple sinks are placed on the boundaries
containing the corners of this area. The network configuration is shown in Fig.4. In the
experiments performed, the value attenuation factor accompanying hop (dr) and the
"value to self" of each sink node in-troduced in our scheme were set to 0.5 and 100.0,
respectively.

Simulation size	2400m × 2400m
Number of sensor nodes	750, 1000, 1250
Range of radio wave	150m or 200m
Number of sinks	2 or 3
Size of each data packet	18 [bytes]
Size of each control packet	6 [bytes]
Battery capacity of each sensor node	0.2 [J] or 0.5[J]

Table 1. Conditions of simulation

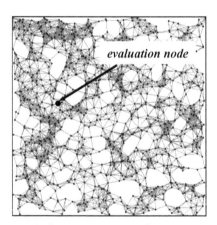

Fig. 4. Large scale and dense wireless sensor network consisting of many static sensor
nodes

In the experimental results reported, our scheme (Matsumoto et al., 2010) is evaluated through a comparison with existing ones (Dubois-Ferriere et al., 2004; Oyman & Ersoy, 2004;
Ohtaki et al., 2006; Utani et al., 2008) where the parameter settings that produced good
results in a preliminary investigation were adopted in preference to existing ones.

3.2 Experimental results on simulation model with two sinks
In this subsection, experimental results on the simulation model with two sinks of our scheme without transmission power control are shown, where the number of sensor nodes was
1000, the range of radio wave and the battery capacity of each sensor node were set to 150m
and 0.5J, respectively.

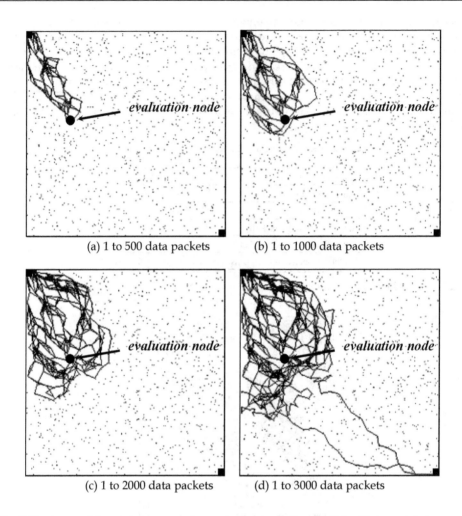

(a) 1 to 500 data packets   (b) 1 to 1000 data packets

(c) 1 to 2000 data packets   (d) 1 to 3000 data packets

Fig. 5. Routes used by applying our scheme to the simulation model with two sinks

As the first experiment on the simulation model with two sinks, it was assumed that the evaluation node marked in Fig.4 detected an abnormal value and transmitted the data packet with this abnormal value periodically. The routes used by applying our scheme are shown in Fig.5. Of the 3000 data packets transmitted from the evaluation node, the routes used by the first 500 data packets are illustrated in Fig.5(a), those used by the 1000 data packets are in Fig.5(b), those used by the 2000 data packets are in Fig.5(c), and those used by a total of 3000 data packets are in Fig.5(d). From Fig.5, it can be confirmed that our scheme enables the autonomous load-balancing transmission of data packets to two sinks using multiple routes.

Next, it was assumed that data packets were periodically transmitted from a total of 20 sens-or nodes placed in the set simulation area. In Fig.6, the transition of the delivery ratio of the total number of data packets transmitted from a total of 20 randomly selected

sensor nodes is shown, and the lifetime of the simulation model with two sinks, as in
Fig.5, is compared. In Fig.6, the existing schemes in Ohtaki et al., 2006 and Utani et al.,
2008, which belong to the category of ant-based routing algorithms, are denoted as *MUAA*
and *AAR*, respectively. The existing scheme in Dubois-Ferriere et al., 2004 and Oyman and
Ersoy, 2004, which describe representative data gathering for a wireless sensor network
with multiple sinks, is denoted as *NS*. From Fig.6, it can be confirmed that our scheme
denoted as *Proposal* in Fig.6 achieves a longer-term operation of a wireless sensor network
with multiple sinks than the existing ones because it improves and balances the load of
each sensor node by the communication load reduction for network control and the
autonomous load-balancing data transmission. Through simulation experiments, it was
verified that our scheme (Matsumoto et al., 2010) is substantially advantageous for the
long-term operation of a large scale and dense wireless sensor network with multiple
sinks.

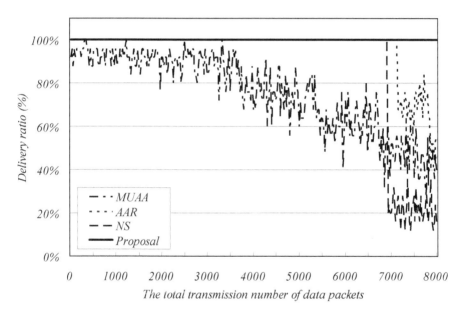

Fig. 6. Transition of delivery ratio

3.3 Experimental results on simulation model with three sinks
In this subsection, through experimental results on the simulation model with three
sinks, the effectiveness of the transmission power control introduced in our scheme is
evaluated. In the following experimental results, the battery capacity of each sensor node
was set to 0.2J, and the range of radio wave of high power transmission in each sensor
node was set to 200 m and it of low power transmission in each sensor node was set to
150m.
As the first experiment on the simulation model with three sinks, it was assumed that the
evaluation node marked in Fig.4 detected an abnormal value and transmitted the data packet with this abnormal value periodically, as in the above subsection 3.2. The routes used by

applying our scheme are shown in Figs.7, 8 and 9, where the number of sensor nodes is 1000. In Figs.7, 8 and 9, T_e was set to 0.0J, $E\times0.5$J, and $E\times0.9$J, where E indicates the battery capaci-ty of each sensor node. Of the 3000 data packets transmitted from the evaluation node, the r-outes used by the first 500 data packets are illustrated in Figs.7, 8 and 9(a), those used by the 1000 data packets are in Figs.7, 8 and 9(b), those used by the 2000 data packets are in Figs.7, 8 and 9(c), and those used by a total of 3000 data packets are in Figs.7, 8 and 9(d). From Figs. 7, 8 and 9, it can be confirmed that the effect of our scheme is extended by early switching to low power.

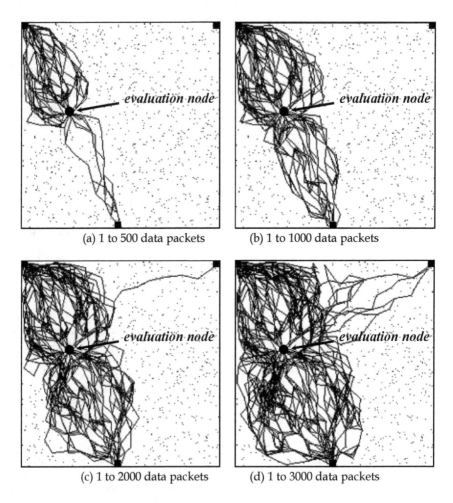

(a) 1 to 500 data packets   (b) 1 to 1000 data packets

(c) 1 to 2000 data packets   (d) 1 to 3000 data packets

Fig. 7. Routes used by applying our scheme (T_e = 0.0J)

Next, it was assumed that data packets were periodically transmitted from a total of 20 sens-or nodes placed in the set simulation area. In Figs.10, 11 and 12, the transition of the delivery ratio of the total number of data packets transmitted from a total of 20 randomly selected sensor nodes is shown, and the lifetime of the simulation model with three sinks, as in Figs.7, 8 and 9, is compared. From Figs.10, 11 and 12, it can be confirmed that the effect of our scheme is extended by early switching to low power in high node density.

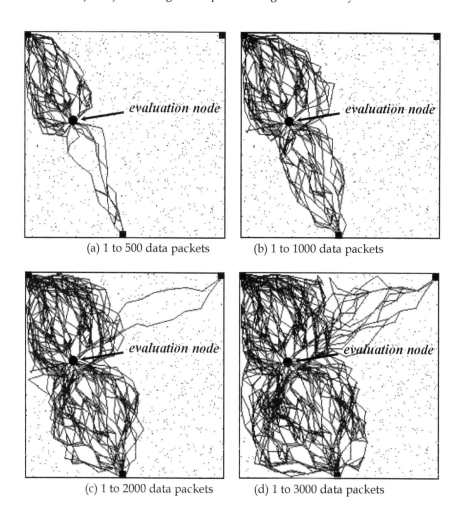

(a) 1 to 500 data packets   (b) 1 to 1000 data packets

(c) 1 to 2000 data packets   (d) 1 to 3000 data packets

Fig. 8. Routes used by applying our scheme ($T_e = E \times 0.5J$)

3.4 Discussion

To facilitate ubiquitous information environments by wireless sensor networks, their control mechanisms should be adapted to the variety of types of communication, depending on ap-plication requirements and the context. Currently, adaptive communication protocols for the long-term operation of the above ubiquitous sensor networks (Intanagonwiwat et al., 20-03; Silva et al., 2004; Heidemann et al., 2003; Krishnamachari & Heidemann, 2003; Wakabay-ashi et al., 2007) are under study. In addition, the advanced design schemes of wireless sens-or networks, such as sink node allocation schemes based on the particle swarm optimization algorithms aiming to minimize total hop counts in a network and to reduce the energy cons-umption of each sensor node (Kumamoto et al., 2008; Yoshimura et al., 2009; Taguchi et al., 2010), and forwarding node set selection schemes (Nagashima et al., 2009; Sasaki et al., 2010) and forwarding power adjustment scheme (Nagashima et al., 2011) for adaptive and efficie-nt query dissemination throughout a wireless sensor network, are positively researched. By coupling our scheme (Matsumoto et al., 2010) with the above advanced design schemes, it can be expected that the lifetime of a wireless sensor network is moreover prolonged.

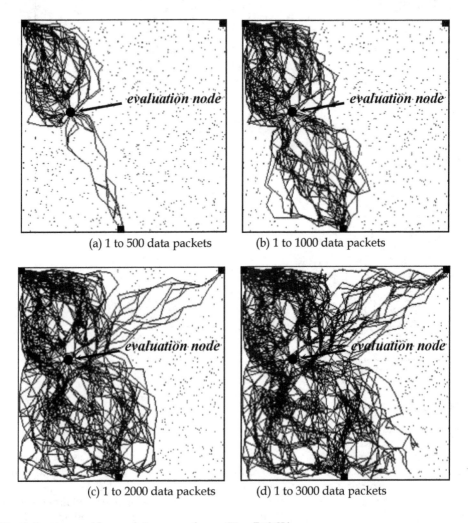

(a) 1 to 500 data packets   (b) 1 to 1000 data packets

(c) 1 to 2000 data packets   (d) 1 to 3000 data packets

Fig. 9. Routes used by applying our scheme ($T_e = E \times 0.9J$)

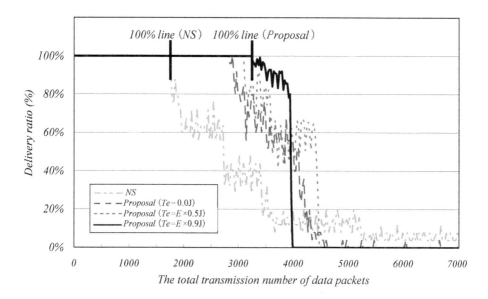

Fig. 10. Transition of delivery ratio (The number of sensor nodes is 750)

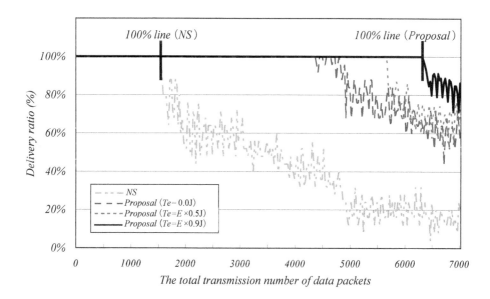

Fig. 11. Transition of delivery ratio (The number of sensor nodes is 1000)

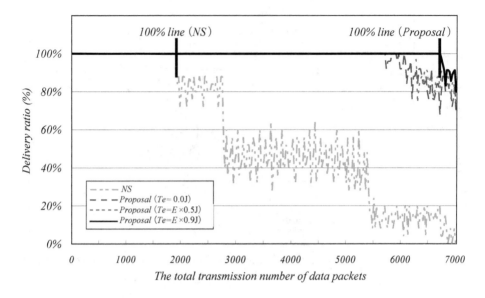

Fig. 12. Transition of delivery ratio (The number of sensor nodes is 1250)

4. Conclusions

In this chapter, a new data gathering scheme with transmission power control that adaptive-ly reduces the load of load-concentrated nodes and facilitates the long-term operation of a large scale and dense wireless sensor network with multiple sinks, which is an autonomous load-balancing data transmission one devised by considering the application environment of a wireless sensor network to be a typical example of complex systems, has been represen-ted. In simulation experiments, the performances of this scheme were compared with those of the existing ones. The experimental results indicate that this scheme is superior to the exi-sting ones and has the development potential as a promising one from the viewpoint of the long-term operation of wireless sensor networks. Future work includes a detailed evaluation of the parameters introduced in this scheme in various network environments.

5. Acknowledgment

The development of a new autonomous decentralized control scheme for the long-term ope-ration of wireless sensor networks with multiple sinks represented in this chapter is suppor-ted by the Grant-in-Aid for Scientific Research (Grant No.21500082) from the Japan Society for the Promotion of Science.

6. References

Akyildiz, I.; Su, W.; Sankarasubramaniam, Y. & Cayirci, E. (2002). Wireless sensor networks: A survey, *Computer Networks Journal*, Vol.38, No.4, 393-422

Clausen, T. & Jaquet, P. (2003). Optimized link state routing protocol, *Request for Comments (RFC) 3626*

Dasgupta, K.; Kalpakis, K. & Namjoshi, P. (2003). An efficient clustering-based heuristic for data gathering and aggregation in sensor networks, *Proceedings of IEEE Wireless Communications and Networking Conference*, 16-20

Dubois-Ferriere, H.; Estrin, D. & Stathopoulos, T. (2004). Efficient and practical query scoping in sensor networks, *Proceedings of IEEE International Conference on Mobile Ad-Hoc and Sensor Systems*, 564-566

Heidemann, J.; Silva, F. & Estrin, D. (2003). Matching data dissemination algorithms to application requirements, *Proceedings of 1st ACM Conference on Embedded Networked Sensor Systems*, 218-229

Heinzelman, W.R.; Chandrakasan, A. & Balakrishnan, H. (2000). Energy-efficient communication protocol for wireless microsensor networks, *Proceedings of Hawaii International Conference on System Sciences*, 3005-3014

Intanagonwiwat, C.; Govindan, R.; Estrin, D.; Heidemann, J. & Silva, F. (2003). Directed diffusion for wireless sensor networking, *ACM/IEEE Transactions on Networking*, Vol.11, 2-16

Jin, Y.; Jo, J. & Kim, Y. (2008). Energy-efficient multi-hop communication scheme in clustered sensor networks, *International Journal of Innovative Computing, Information and Control*, Vol.4, No.7, 1741-1749

Johnson, D.B.; Maltz, D.A.; Hu, Y.C. & Jetcheva, J.G.(2003). The dynamic source routing protocol for mobile ad hoc networks, *IETF Internet Draft*, draft-ietf-manet-dsr-09.txt

Krishnamachari, B. & Heidemann, J. (2003). Application-specific modeling of information routing in wireless sensor networks, *Technical Report*, ISI-TR-2003-576, USC-ISI

Kumamoto, A.; Utani, A. & Yamamoto, H. (2008). Improved particle swarm optimization for locating relay-dedicated nodes in wireless sensor networks, *Proceedings of 2008 Joint 4th International Conference on Soft Computing and Intelligent Systems and 9th International Symposium on Advanced Intelligent Systems*, 1971-1976

Matsumoto, K.; Utani, A. & Yamamoto, H. (2009). Adaptive and efficient routing algorithm for mobile ad-hoc sensor networks, *ICIC Express Letters*, Vol.3, No.3(B), 825-832

Matsumoto, K.; Utani, A. & Yamamoto, H. (2010). Bio-inspired data transmission scheme to multiple sinks for the long-term operation of wireless sensor networks, *International Journal of Artificial Life and Robotics*, Vol.15, No.2, 189-194

Nagashima, J.; Utani, A. & Yamamoto, H. (2009). Efficient flooding method using discrete particle swarm optimization for long-term operation of sensor networks, *ICIC Express Letters*, Vol.3, No.3(B), 833-840

Nagashima, J.; Utani, A. & Yamamoto, H. (2011). A study on efficient query dissemination in distributed sensor networks -Forwarding power adjustment of each sensor node using particle swarm optimization-, *Proceedings of 16th International Symposium on Artificial Life and Robotics*, 703-706

Nakano, H.; Utani, A.; Miyauchi, A. & Yamamoto, H. (2009). Data gathering scheme using chaotic pulse-coupled neural networks for wireless sensor networks, *IEICE Transactions on Fundamentals*, Vol.E92-A, No.2, 459-466

Nakano, H.; Utani, A.; Miyauchi, A. & Yamamoto, H. (2011). Chaos synchronization-based data transmission scheme in multiple sink wireless sensor networks, *International Journal of Innovative Computing, Information and Control*, Vol.7, No.4, 1983-1994

Ogier, R.; Lewis, M. & Templin, F.(2003). Topology dissemination based on reverse-path forwarding (TBRPF), *IETF Internet Draft*, draft-ietf-manet-tbrpf-10.txt

Ohtaki, Y.; Wakamiya, N.; Murata, M. & Imase, M. (2006). Scalable and efficient ant-based
routing algorithm for ad-hoc networks, *IEICE Transactions on Communications*, Vol.E
89-B, No.4, 1231-1238

Oyman, E.I. & Ersoy, C. (2004). Multiple sink network design problem in large scale wireless
sensor networks, *Proceedings of 2004 International Conference on Communications*, Vol.
6, 3663-3667

Perkins, C.E. & Royer, E.M. (1999). Ad hoc on-demand distance vector routing, *Proceedings of
2nd IEEE Workshop on Mobile Computing Systems and Applications*, 90-100

Rajagopalan, R. & Varshney, P.K. (2006). Data aggregation techniques in sensor networks: A
survey, *IEEE Communications Surveys and Tutorials*, Vol.8, 48-63

Sasaki, T.; Nakano, H.; Utani, A.; Miyauchi, A. & Yamamoto, H. (2010). An adaptive selection scheme of forwarding nodes in wireless sensor networks using a chaotic neural
network, *ICIC Express Letters*, Vol.4, No.5(A), 1649-1655

Silva, F.; Heidemann, J.; Govindan, R. & Estrin, D. (2004). Directed diffusion, *Technical
Report*, ISI-TR-2004-586, USC-ISI

Subramanian, D.; Druschel, P. & Chen, J. (1998). Ants and reinforcement learning: A case study in routing in dynamic networks, *Technical Report TR96-259*, Rice University

Taguchi, Y.; Nakano, H.; Utani, A.; Miyauchi, A. & Yamamoto, H. (2010). A competitive particle swarm optimization for finding plural acceptable solutions, *ICIC Express Letters*, Vol.4, No.5(B), 1899-1904

Utani, A.; Orito, E.; Kumamoto, A. & Yamamoto, H. (2008). An advanced ant-based routing
algorithm for large-scale mobile ad-hoc sensor networks, *Transactions on SICE*, Vol.
44, No.4, 351-360

Wakabayashi, M.; Tada, H.; Wakamiya, N.; Murata, M. & Imase, M. (2007). Proposal and evaluation of a bio-inspired adaptive communication protocol for sensor networks, *I-
EICE Technical Report*, Vol.107, No.294, 89-94

Wakamiya, N. & Murata, M.(2005). Synchronization-based data gathering scheme for sensor
networks, *IEICE Transactions on Communications*, Vol.E88-B, No.3, 873-881

Xia, L.; Chen, X. & Guan, X.(2004). A new gradient-based routing protocol in wireless sensor
networks, *Lecture Notes in Computer Science*, Vol.3605, 318-325

Yamamoto, I.; Ogasawara, K.; Ohta, T. & Kakuda, Y. (2009). A hierarchical multicast routing
using inter-cluster group mesh structure for mobile ad hoc networks, *IEICE Transactions on Communications*, Vol.E92-B, No.1, 114-125

Yoshimura, M.; Nakano, H.; Utani, A.; Miyauchi, A. & Yamamoto, H. (2009). An effective allocation scheme for sink nodes in wireless sensor networks using suppression PSO,
ICIC Express Letters, Vol.3, No.3(A), 519-524

6

Environmental Monitoring WSN

Ittipong Khemapech

University of the Thai Chamber of Commerce,
Thailand

1. Introduction

Energy conservation is currently growing in importance. This chapter focuses on the issue of energy conservation within the domain of Wireless Sensor Network (WSN). There are also specific reasons why energy conservation is more important for WSN than for other types of networks. A WSN consists of multiple sensors which are able to sense some aspect of their environment and communicate their readings to a base station or sink without being physically connected to it. Sensors are often also resource constrained, being small in size and relying on small batteries for power. Consequently, the efficient utilisation of energy should be an important priority for designing WSN network protocols. This differs from the traditional approach to designing network protocols where issues like survivability, maximising throughput or reliability have been prioritised. Making energy conservation an important design priority is a new approach.

Wireless sensor network (WSN) is an important research area with a major technological impact. With significant breakthroughs in "Micro Electromechanical Systems", or MEMS, (Warneke & Pister, 2002), sensors are becoming smaller. It is feasible to fit them into a smaller volume with more power and less production costs. WSN may be deployed in a wide range of different environments. These include remote and hostile environments as well as local and friendly ones. A major driving force behind research in WSN has been military and surveillance applications. Recently, however diversification has occurred with the development of civil applications. One example which is used as a reference point throughout this work is Great Duck Island (GDI). Sensors were scattered over a remote island to monitor the seabird's migration (Mainwaring et al., 2002). In another example WSN was deployed around volcanoes (Allen et al., 2006). Such applications illustrate the usefulness of WSN which make data collection feasible from remote and hostile environments with minimal human intervention.

One of the main objectives of WSN power conservation is to minimise energy usage whilst other functional requirements such as reliability or time synchronisation are still achieved. Some authors argue that multi hop communication allows for deployment in scenarios where direct communication with a base station is not practical (Arora et al., 2004; Allen et al., 2006; Chintalapudi et al., 2006). However, the spread of the Internet means that wireless devices may often communicate directly with a device that is connected to the Internet and has a reliable power supply. This work focuses on the design of wireless sensor networks protocols where direct communication with a powered base station is feasible and data is sent from the sensors to the base station at regular intervals. There are several important scenarios where such two assumptions hold.

This research work specifically looks at WSN where direct communication is possible and beneficial. A protocol for WSN, Power & Reliability Aware Protocol (PoRAP), is developed and provides energy efficient data delivery, without increasing packet loss. In designing PoRAP several experiments were conducted to establish the relationship between transmission power, reception signal strength and packet reception success. These showed a strong correlation between Received Signal Strength Indicator (RSSI) and Packet Reception Rate (PRR). In PoRAP, the RSSI is monitored at the base station. If the RSSI is too high the base station signals the sensor to reduce its transmission level, thereby saving power. If the RSSI is too low the base station signals the sensor to increase its transmission level so that packet loss is avoided.

PoRAP adopts a schedule based scheme for the sources' transmissions. It is assumed that nodes will be reporting measurement data regularly back to the base station. Each reporting interval consists of three time periods. In the first the base station sends a configuration packet. This informs nodes whether they are to increase, decrease or leave unaltered their transmission levels. There are then slots, each of which contains a data slot within which a sensor may transmit its data to the base station. There may then be a period of quiet before the start of a new cycle. Delays and clock drifts are measured so that nodes know when to wake up to listen and transmit. Delays depend upon payload size.

The design aims to optimise energy conservation rather than system throughput, in many sensing scenarios high throughput is not required. Sensors collect and report some physical data such as temperature and humidity. In such cases, the data reporting rate may be in minutes or hours. Two packet structures are used in PoRAP. The control packet is used in control and setup phase. It contains essential information for transmission power adaptation and scheduling. The data packet is used to deliver the collected physical data back to the base station.

The remaining parts of this chapter is organised as follows: Section 2 addressed application specific WSN. At present, WSN has been used in both military and civil applications. Each application category has particular characteristics and its own set of requirements. Hence, there are significant challenges in a generic protocol design for a variety of applications. Resource constraint issues are provided in Section 3. Apart from limited power resources, sensors also have constrained communication ranges for indoor and outdoor environments. The distance between the source and destination is crucial to employing an appropriate underlying communication paradigm. Section 4 describes the experimental details and their results which motivate the design of PoRAP. There are several factors which affect the link quality metrics such as distance between source and base station and time of day. The design of PoRAP is outlined in Section 5. PoRAP consists of several TinyOS components at the source and base station. The results shown in Section 4 motivates the design. The results of PoRAP evaluation in terms of energy conservation are presented in Section 6. Lower transmission power can be used to save the power whilst the reliability is in the desired range. Finally, Section 7 concludes the chapter.

2. Application specific WSN

Apart from being used in military or surveillance, WSN has been deployed in several civil applications which have different requirements. Periodic sensing is required in some habitat and environmental monitoring systems whilst event sensing is the norm in surveillance systems. Network lifetime and data reporting rates are therefore major concerns for the former and latter cases, respectively. To be application specific results in a more complicated design process, especially in the case of designing a generic power-aware protocol.

In total, seven groups of applications have been categorised by us based upon their functionalities including habitat monitoring (HM) (Juang et al., 2002; Mainwaring et al., 2002; Szewczyk et al., 2004), environmental monitoring (EM) (Allen et al., 2006; Martinez et al., 2005), health monitoring (HEM) (Jovanov et al., 2003, Otto et al., 2006), structural health monitoring (SHM) (Chintalapudi et al., 2006; Kottapalli et al., 2003; Paek et al., 2005, Schmid et al., 2005), event detection and tracking (EDT) (Arora et al., 2004; Dreicer et al., 2002; Simon et al., 2004), transport monitoring (TM) (Coleri et al., 2004) and location-aware system (LAS) (Brignone et al., 2005). Specific capabilities and underlying communication paradigms have been outlined. For example, data encryption may be required in some health monitoring systems for transmitting a patient's diagnosis data to the main server located at the hospital. Furthermore, data correctness is also required in this case. In some applications such as event tracking and detection systems, several intermediate nodes are required for forwarding the sensed data to the base station. However, a direct communication from source to base station is found in some health monitoring systems. This section addresses application specific characteristics of WSN applications by detailing the differences in their requirements.

2.1 Event/periodic based

The "Event/Periodic Based" aspect demonstrates how often data reporting is conducted. There are three main types including event-based, periodic-based and hybrid. Each sensor is triggered to operate by the occurrence of an event in the case of an event-based application. An example of this application type is the Event Detection and Tracking. Congestion is one of the major concerns designing a protocol to support event-based networking as it is caused by a lot of traffic generated by all sources in an event area. The key idea of congestion avoidance is to control data reporting rate of such sensors (Sankarasubramaniam et al., 2003). However, the main assumption is that all data packets have the same priority. Packet loss is therefore tolerantly acceptable. There are several works on congestion control specifically developed for WSN (Ee & Bajcsy, 2004; Hull et al., 2004, Lu et al., 2002, Wan et al., 2003). The congestion control approach focuses on channel monitoring to dynamically adjust the data forwarding rate. CODA (Wan et al., 2003) has been designed to cover two types of problems corresponding to the deployed sensors and their data rate. However, it does not provide any queue occupancy monitoring. Sending an ACK (Acknowledgement) in the case of persistent congestion, even if it is small in size, may increase the number of traffic. This mechanism also requires feedback signalling which results in higher cost. Only packet prioritisation could be found in (Lu et al., 2002). However, it proposes the VMS (Velocity Monotonic Scheduling) policy which supports both static and dynamic computation of the requested velocity and it also solves the fairness problem. Both channel and queue occupancy monitoring are provided in (Hull et al., 2004) and (Ee & Bajcsy, 2004). A child node can transmit packets only when its parent does not experience congestion problems and some help from the MAC (Medium Access Control) layer to shift the transmitting time to avoid interference are proposed in (Hull et al., 2004). A similar concept also exists in (Ee & Bajcsy, 2004) by comparing the normalised rate of a node and its parents.

Each sensor periodically performs its operation. Some examples of data collected by the sensors are temperature and humidity. The significant change in readings may be used to identify the presence of seabirds (Mainwaring et al., 2002) and intruders (Arora et al., 2004). Instead of heavily generated traffics, both sensor and network lifetimes are the core requirement of this application type. Finally, both event and periodic sensing operations may be desired in some applications such as SHM (Structural Health Monitoring) and EDT systems. For example, the displacement of construction elements is periodically reported for maintenance purposes whilst an event-based operation is applied for warning and evacuating notifications during an earthquake.

This work focuses on developing a power-aware protocol which supports an efficient data delivery in periodic based applications such as health, habitat and environmental monitoring where the data reporting rate is in minutes or hours. Sensors may be scattered over a remote and hostile area to collect and report physical data and they should have to operate for months. Hence, battery lifetime is important and one of the main goals is to conserve communication energy.

2.2 Mobility of sources

The mobility of sources or sensors can be found in some particular applications such as HM (Habitat Monitoring, HEM (HEalth Monitoring and LAS (Location-Aware System). In some cases, sensors are attached to the targeted objects or location (Jovanov et al., 2003; Juang et al., 2002, Martinez et al., 2005) in order to monitor the data of interest or current location. Mobile sensor networks have a different set of supporting infrastructures compared to the traditional WSN. It is essential for each mobile sensor to know its own location. The GPS (Global Positioning System) is used for locating sensors which are attached to the goods. Alternatively, several nodes with known locations may be used as references for the others to calculate their own locations [Brignone et al., 2005]. The issues of sensor mobility are beyond the scope of this work.

2.3 Mobility of sources

Wireless sensor network (WSN) consists of sensors which are wirelessly connected. The main objective of WSN development is to collect physical data from an area of interest. Therefore, communication between sensors is a key aspect. Normally there are two node types in WSN including the source and base station. Sources are ordinary sensors having limited resources whereas base stations are assumed to have more power and other resources. The main duty of sensors is collecting and transmitting data to the destination or base station. The sensors are probably required to cover a large area and direct communication between sources and base station is unlikely due to limited communication range. Several intermediate sensors responsible for forwarding data packets to the base station are therefore required. This is known as multi-hop communication. Each sensor also acts as a routing node in order to find the shortest or cheapest path by means of power consumption. Several applications deploy multi-hop communication (Allen et al., 2006; Chintalapudi et al., 2006; Schmid et al., 2005; Dreicer et al., 2002; Simon et al., 2004). The multi-hop approach has several advantages. For example, a new path is discovered when some sensors die. Deploying a large number of cheap sensors over a large area is feasible as the sensors can act as routing nodes and the collected data is forwarded to the destination. However, one of its drawbacks is each node has to listen to the channel most of the time in order to detect if a message is arriving. The sensors have to conduct some computations in order to discover the cheapest path. Moreover, communication with its neighbours is another requirement to set up a selected path. Such processes require a significant amount of power, taken from the battery power available. Introducing several intelligent features to each sensor is also limited due to the power constraint.

Each source can transmit the data directly to the base station if the sources are located within the base station's communication range. Some examples of existing applications deploying single-hop communication (Mainwaring et al., 2002; Martinez et al., 2005; Jovanov et al., 2003; Otto et al., 2006). For single-hop, the sources are located within the base station's range. Direct communication is therefore feasible and several benefits are realised. One of the advantages is the ability to introduce a variety of intelligent features to the base station as it is assumed to have more power and computational capabilities compared to an ordinary sensor. Each source does not require the power necessary for routing. Idle listening can be minimised as the sources can be switched to sleep mode if they do not transmit data or receive the control packet. The base station controls the communication schedule of its sources to avoid data collisions. Power for carrier sensing is not desired. In multi-hop, each source is responsible for sensing, data reporting and routing. The number of transmissions and receptions increases with the number of intermediary nodes required for data forwarding.

This work looks at protocol development for single-hop. A scenario where the single-hop is viable is Environmental Monitoring (EM). Sources and base stations are distributed and several clusters or patches are formed. A power-aware, single-hop protocol can thus be used in each of the clusters (Mainwaring et al., 2002). A low duty cycle is the norm in EM so the communication cycle of each source can be scheduled by the base station. A time slot is allocated to each source to perform data transmissions. Carrier sensing is thus not required in this scheme. The sources synchronise to the base station by checking the information included in the control packet.

2.4 Reliability

Wireless sensor network (WSN) has been currently deployed in several civil applications. The physical data is collected and transmitted for further analysis. The issue of reliability in data delivery is important for providing complete reliability consumes a significant proportion of power. Applying the Transmission Control Protocol (TCP) protocol to WSN is expensive because of its three-way handshake mechanism and packet header size. The User Datagram Protocol (UDP) is considered to be more suitable for sensors although it was designed to provide unreliable data transport. In some applications, data loss may be not a serious problem because of the large amount of deployed sensors. Reliable data transport is important for some types of data such as control messages delivered by the base station (Wan et al., 2002). The following paragraphs provide some details of reliable transport protocol for WSN researches including PSFQ (Pump Slowly, Fetch Quickly) (Wan et al., 2002), ESRT (Event-to-Sink Reliable Transport) (Sankarasubramaniam et al., 2003), and RMST (Reliable Multi-Segment Transport) (Stann & Heidemann, 2003).

One of the main goals to achieve reliable data transport is to orchestrate data receiving and forwarding processes to lessen the packet loss due to buffer overflow. PSFQ proposes three different operations including pump, fetch and report. Data generated from a source node is injected slowly into the network in order to allow such nodes experiencing data loss to fetch the missing packets very aggressively. Timing is a core process in order to avoid operational synchronisation. A hop-by-hop recovery is applied to avoid exponential error accumulation as occurs in the end-to-end scheme. Data delivery status information can be sent back to users or applications in a piggyback fashion.

Focusing only on the forward or sensor-to-sink direction, ESRT was designed to provide a reliable data transport by inspecting current network state in terms of reliability and congestion. The state result is categorised and the reporting frequency is then repetitively adjusted to reach an optimal point. ESRT provides both reliable data transport and congestion control. Local buffer level monitoring is used to detect congestion.

Directed Diffusion (Intanagonwiwat et al., 2003) is a routing protocol which provides a multipoint-to-multipoint communication. A sink firstly indicates an interest and propagates it to the nodes. Interest and node information is kept as gradients. The optimised reinforced path is then established to send the attribute-value pairs data. RMST is implemented as a filter to provide some information about the data fragment such as ID and total number of fragments to detect loss. A NACK (Negative ACKnowledgement) is sent via a back-channel to upstream neighbouring nodes in case of data loss.

According to the above fundamental protocol descriptions, several conclusions can be made. In a densely deployed environment, data loss may be accepted. However, this condition may apply only in the case of sensor-to-sink traffic. The sink or base station plays a major role in the network by broadcasting several control packets to the sensors. Such packets should not be lost. Moreover, there are various types of sensing data, such as structural displacement due to wind or earthquake (Xu et al., 2004), which need some combination from different nodes to create usable data before forwarding that data to the sink. PSFQ designing concepts are more complicated but can be applied to a broader area of application. The data retransmission mechanisms are not mentioned in ESRT as it focuses on statistical reliability. However, PSFQ does not provide congestion control schemes as ESRT does. RMST is designed to run over the Directed Diffusion routing protocol. Although it may take the least effort compared to the other two, it is not generic enough.

3. Resource constraint issues

This section introduces several issues of resource constraint in WSN. A sensor can be considered as a small computing device which is capable of sensing, data processing, storage and communication. Sensors are deployed in an area of interest and they may have to operate without maintenance throughout their lifetimes. Power is thus one of the limited resources. Unless an external source of energy is provided, power for all operations comes from batteries. Two AA batteries are required in the widely used platforms such as Tmote, Telos and Mica. The capacity of the AA battery is approximately 2,000 to 3,000 milli-ampere-hour (mAh). In order to calculate the battery life, the capacity is divided by the actual load current and the obtained lifetime is in hours. An equation for calculating sensor's lifetime is given in (Polastre et al., 2004) where the lifetime is equal to the product between capacity (mAh) and voltage (3V) divided by total energy consumption in micro-joules. The default capacity defined in (Polastre et al., 2004) is set at 2,500mAh.

Communication accounts for a significant proportion of energy consumption. There are four main modes of communication including sending, receiving, sleeping and listening. The transceiver is one of the major sensor components and it makes them capable of communicating with other nodes. Recent transceivers or radio chips such as CC1000 and CC2420 provide programmable transmission power. Sensors consume less power when they send at a lower power level. Hence, transmission power control is one of power-aware schemes in WSN. The sensors do not always send at the maximum power. Tmote platform is chosen in this study and it employs CC2420 transceiver. For the CC2420 mote the minimum and maximum transmission power is 8.5 and 17.4 milli-amperes (mA). Over 50% of the power can be saved if the minimum power is always used.

Sensors equipped with CC2420 radio chips consume a greater amount of power when they receive data. According to the data sheet, 19.7mA is required for reception. Listening and sleeping consume 365 and 20 micro-amperes (μA), respectively. Hence, in the case of the CC2420 mote, data reception consumes more energy than transmission. The base station is the destination and it may be connected to a desktop or laptop computer. In such cases, the base station has extra power from the connected machine. However, the sensors which act as intermediary nodes between source and destination have to receive and forward packets resulting in sensor's lifetimes being decreased. The listening power is approximately 17 times greater than sleeping. In some applications such as environmental monitoring, the data sampling interval may be in minutes or hours. The transceivers should be switched to sleep mode instead of listening. Scheduling issues occur when two nodes communicate with each other. The data is not received if the receiver is in sleep mode. The nodes have to agree upon the same scheduling. Another scheme based upon contention-based can be used; the receiver can periodically listen to the signal propagated over the medium to inspect whether the incoming message is destined for it.

WSN is also a shared medium system. Each of the sources and base station has to engage the medium to perform data communication. Data collisions occur if the sources transmit at the same time and energy will be wasted by unsuccessful data delivery. A Medium Access Control (MAC) protocol is required to resolve the contention. The features of the MAC protocol together with the application behaviour determine when a node is idle, when it is listening and when it is sending. As each of these states have different power requirements the MAC protocol impacts upon the efficiency of operation and the power consumption. There are two main MAC schemes; the contention and the schedule based. The medium is sensed prior to transmission and the sensors have to backoff if the medium is declared busy. This work focuses on the single-hop where the sources send data directly to the base station. Another scheme, schedule based, is adopted. A data slot is allocated to each node. No carrier sensing and corresponding energy is required. The main issue is that the slot must be long enough for completing data delivery, otherwise, data collisions are likely. Experimentations required to determine the duration required for both sending and receiving together with the effective factors such as data payload size. Each node is switched to sleep mode to spend the least amount of power when its slot does not arrive.

The buffering capacity of CC2420 is limited to 128 bytes. Taking the header's and footer's sizes into account, the allowable data payload size is thus less than 128 bytes. Apart from sensed data, some control information is required in the packet such as identification and timestamp. Additional packet structures may be required if all the information cannot be stored in one packet. Control overhead is considered as one of the costs and should be minimised in order to decrease transmission and reception energy.

Wireless sensor network (WSN) has been currently deployed in several surveillance and civil applications. Sensors may be scattered over an area of interest which can be very large. The communication range is thus important and depends upon the selected transceiver. For example, the CC2420 mote has 50m and 125m indoor and outdoor ranges. Under some circumstances, the maximum transmission power may not produce the maximum ranges. Furthermore, sending data to the node located at farther distances requires higher transmission power. Multi-hop is therefore usually used in WSN. Several intermediary sensors are required for data forwarding from the source to destination. Single-hop communication is feasible if the destination is located within the source's range. Multiple transmissions and receptions are not required if direct communication applies. However, the same transmission power cannot always be used as the link quality changes over time. The next section describes several sources of variability in radio frequency

4. Motivation of PoRAP development

This work aims at building a communication protocol for WSN. The targeted scenario is the periodic-based where a low duty cycle is required. The network consists of a fixed set of sources and a base station. Furthermore, direct data communications between the base station and its sources are feasible. The communication protocol to be developed will effectively support the single-hop WSN. Such a structure forms a network cluster which can be used in some environmental or habitat monitoring system such as (Mainwaring et al., 2002) and (Tolle et al., 2005). As the number of sources is fixed throughout the communications, the data reporting rate is fairly constant. The communication of the sources can be therefore scheduled and controlled by the base station. A time slot is allocated to each source and will be used for data communication. Only one source can use the shared medium whilst the others switch to sleep mode by turning their radios off and consuming the least amount of energy. Data collision can be avoided and idle listening can be minimised.

4.1 Sensor node power consumption

This section establishes the significance of network communication as a consumer of energy within a wireless sensor network. In doing so a careful reading of sensor data sheets is used to inform calculations based upon the sensor's parameters and simulations. What proportion of the power is used for communication is investigated and how power may be conserved is identified.

In order to investigate how power is consumed by a sensor, a simulation study has been established. The results are validated by the CC1000 transceiver data sheet. As the sensor operating system used in this work is TinyOS, the selected simulator is TOSSIM which is a TinyOS library. TinyOS is an operating system specifically designed for embedded devices such as sensors. It has been widely used in both research and commercial communities. The selected release of the simulator is TOSSIM 1 and it does not provide power usage measurement capability. PowerTOSSIM, an extension module developed for analysing power consumption of hardware components (Shnayder et al., 2004) is used to address the investigation on power consumption and it is included in Tiny 1.1.11. The only sensor platform supported in PowerTOSSIM is Mica2 which employed the CC1000 radio chip. The PowerTOSSIM supports an operating frequency of 400 Megahertz (MHz) and a voltage of 3 Volt. The energy model file of PowerTOSSIM adopts the required transmission current for each power level. According to the CC1000 datasheet, 31 output power levels ranging from -20 to +10dBm can be programmed. The dBm is the measurement of power loss in decibels (dB) using 1 milli-watt (mW) as a reference value.

4.1.1 Simulation parameters

A sensor node was created in the simulation and performs as a transmitting node. An experiment was conducted to obtain the current consumption required by each transmission power level. In total five transmission powers including -20, -10, 0, +6 and +10dBm were used. The corresponding current consumption was measured by (Shnayder et al., 2004) and their results are shown in Table 1. A simulation duration of 60 seconds and a total of 30 runs were conducted at each power level. A higher current will be consumed if the sensor transmits at a higher power.

Transmission Power (dBm)	Required Current (mA)
-20	5.21
-10	6.10
0	8.47
+6	13.77
+10	21.48

Table 1. Current consumption measured by Shnayder et al., 2004

The results shown in Table 1 were used to compute the energy consumption required by each transmission power level. Fig. 1 shows error-bar plots of radio and total energy consumption at -20, -10, 0, +6 and +10 dBm. An analysis of power usage and conservation with respect to the maximum power level is described in Table 2.

According to Fig. 1, several observations can be made. Firstly, an increase in transmission power results in a higher energy consumption. Transmitting data at lower power uses less energy. For example, over 75% of energy can be conserved if the minimum power is used for transmission instead of the maximum. Secondly, the radio unit consumes a significant amount of energy. Up to 56% and 84% of energy are used by the radio if the sensor transmits at minimum and maximum power levels, respectively. The results are validated by the CC1000 data sheet which is the employed radio in Mica2. According to the CC1000 datasheet, the required current consumption for -20 and +10 dBm are 6.9 and 26.7 milli-amp (mA), respectively. Therefore, over 74% can be conserved and this is close to the 75% which is obtained from PowerTOSSIM.

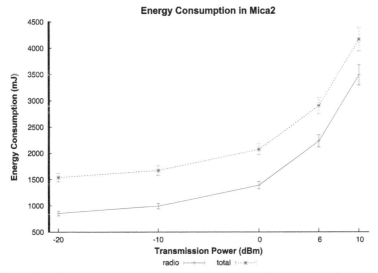

Fig. 1. Radio and total energy consumption at various transmission power levels

Transmission Power (dBm)	Average of Radio Power Consumption (mJ)	Percentage of Used Power	Percentage of Saved Power
-20	861.52	24.67	75.33
-10	1000.33	28.64	71.36
0	1396.44	39.98	60.02
+6	2236.90	64.05	35.95
+10	3492.48	100	0

Table 2. Average radio power consumption (mJ) and percentages of used and saved power

Two key motivations are established with respect to the simulation results. Firstly, transmission power considerably affects radio power consumption. The power-aware approach based upon power adaptation is Transmission Power Control (TPC). PoRAP adopts the TPC concepts in order to achieve the power conservation goal. The selected sensor platform in this work is Tmote and it employs the CC2420 radio instead of the CC1000. Like the CC1000, the CC2420 also supports transmission power adaptation but it provides a different range of power levels. Table 3 shows some of the possible power levels and the corresponding current consumption. An analysis of power conservation with respect to the maximum level is also shown.

Transmission Power (dBm)	Current Consumption (mA)	Percentage of Used Current	Percentage of Saved Current
-25	8.5	48.85	51.15
-15	9.9	56.90	43.10
-10	11.2	64.37	35.63
-7	12.5	71.84	28.16
-5	13.9	79.89	20.11
-3	15.2	87.36	12.64
-1	16.5	94.83	5.17
0	17.4	100	0

Table 3. Transmission power levels provided by CC2420 and analysis of power conservation

According to Table 3, over 50% of power can be saved if the minimum power is used for data transmission. The transmission power is one of the main factors which produces different reception strengths. The power adaptation is based upon the current link quality in order to maintain a good link. However, power adaptation is based upon several factors affecting link quality such as distance and time-of-day.

Secondly, according to Fig. 1, the radio unit accounts for a significant amount of power compared to the total consumed by all hardware components. Keeping the radio in sleep mode after the sensor has transmitted the data may establish an enhancement in power conservation. This is feasible if the single-hop network sensors do not listen to transmissions from other nodes in order to discover optimal data paths. The schedule-based MAC (Medium Access Control) approach suits the direct communication scenario as each of the sources wake up for control reception and data transmission. Otherwise, they are in sleep mode and consume the least amount of communication energy.

4.2 Environmental investigation of transmission power and reliability

This section provides details of experimental studies aimed at establishing effects of transmission power, distances and time-of-day on link quality metrics. In total three metrics including RSSI (Received Signal Strength Indicator), LQI (Link Quality Indication) and PRR (Packet Reception Rate) are used to describe the effects. The relationships between the metrics are also investigated and will be used for establishing power adaptation policies.

4.2.1 Link quality metrics

There is a variety of sources which cause variability in link quality in wireless communication. Unlike wired communication, environmental factors such as climatic conditions and time-of-day also affect the degree of signal attenuation. A significant degree of signal attenuation or interference may lead to unsuccessful data transmission. Link quality measurement is therefore one of the major issues in wireless network communication.

A transmitter sends data packets at a specific transmission power wirelessly over a medium to a receiver. The transmission power level is programmable and this capability is provided by a transceiver or radio unit which is a component responsible for data transmission and reception. A sensor communicates with the other node by sending and receiving messages via wireless channel which is normally air. Several signals are generated from various sources such as electronic appliances and they are dissipated to the air. A wireless channel may then have background noise which is capable of interfering with data delivery between a pair of nodes. Moreover, time-of-day and climatic conditions such as fog and rain affects the wireless link quality. In order to determine link quality characteristics, all causes of signal strength reduction are considered as sources of signal attenuation. The reduced magnitude in signal strength is therefore defined as signal attenuation. If the transmission power is less than signal attenuation, the message cannot be successfully received. When the receiver is not able to receive the sent packet and the number of received packets is not increased, the reliability requirement defined by an application may not be met. Transmission power should be adjusted in response to the changing link quality.

A radio unit provides several mechanisms to measure received signal power. The measured values are categorised as received signal strength (RSS). In total two attributes including RSSI (Received Signal Strength Indicator) and LQI (Link Quality Indication) are in the RSS category. The RSS can be used to indicate link quality. The reliability requirement specified by an application indicates a required number of packets received at the base station. The percentage of data receptions can be used to describe the link quality. The packet reception rate (PRR) is therefore introduced. Relationships amongst transmission power (TX), received signal strength (RSS) based attributes and PRR is useful for mapping application requirements to link quality measurements. Thus, the transmission power is adapted in order to provide reliability of packet reception.

Received Signal Strength Indicator (RSSI) is defined as a measurement of the signal strength of an incoming message. The transmitted signal strength or transmission power reduces as the signal propagates through the medium. The RSSI is measured at the receiver and it demonstrates the received signal strength. Therefore, signal attenuation is approximately the difference between the transmission power and the RSSI. Link Quality Indication (LQI) is another metric in the RSS-based category. According to the definition outlined in IEEE 802.15.4 Standard for Local and Metropolitan Area Networks, the LQI measurement is a characterisation of the strength and/or quality of received packet. Each received packet has its own LQI measurement and the integer value ranges from 0 to 255. Therefore, the minimum and maximum values of LQI for each packet are 0 and 255, respectively. The IEEE standard recommends at least eight unique values of LQI should be used in order to yield a uniform distribution between the two limits. The following details of LQI are based upon the CC2420 radio unit as it is used in both Tmote Sky and Tmote Invent which are the chosen platforms in this research. Apart from RSSI and LQI, PoRAP determines an additional link quality index. The main reason is that both RSSI and LQI are not transparent to the user or application. Mapping mechanisms are required in order to convert an application requirement to the ranges of RSSI and LQI values the base station should have. This subsection aims to describe the Packet Reception Rate (PRR) which is more closely related to the application requirement. In this research, the PRR is defined as a percentage of the number of correctly received to that of transmitted packets. The PRR value is in the range of 0% to 100%. The 100% PRR indicates complete reliability. Each received packet has its own measured RSSI and LQI which can be used to predict the PRR. Models representing relationships amongst metrics are therefore required and demonstrated later in this chapter.

4.2.2 Experimental setup

In our implementation-based experiments, Tmote Invent and Tmote Sky are used as the sensor and base station, respectively. Both of them employ the CC2420 radio which has working frequency band from 2,400 to 2,483 Megahertz (MHz). The radio transmission data rate is 250 kilobits per second (kbps). The random access memory (RAM) and program flash sizes are 10 and 48 kilobytes (Kbytes). The main difference between both platforms is that the Tmote Invent provides built-in sensor and battery boards. The minimum and maximum transmission power levels are -25 and 0dBm, respectively. Tmote sensors consume 8.5 and 17.4 milli-amps (mA) for transmitting a data packet at minimum and maximum power levels, respectively. A current of 19.7mA is required for radio receiving. This indicates that receiving accounts for a large radio power usage. Listening removal in PoRAP may enhance power conservation in WSN. Each Tmote sensor includes an internal Inverted-F antenna which is a wire monopole. The top section of the antenna is folded down to be parallel with the ground plane. The communication ranges for indoor and outdoor are 50m and 125m, respectively.

The experiments were conducted in the 16m x 20m indoor environment. The base station was plugged into a desktop computer and received data from sensors. Three sensors were used and they were placed at the same locations. In total 10 locations including 1, 2, 3, 4, 5, 7, 10, 13, 16 and 20m were used. The sensors and base station had the same antenna orientation and height above floor level. The payload size was 12 bytes. In total 8 transmission power levels including 3, 7, 11, 15, 19, 23, 27 and 31 associated to -25, -15, -10, -7, -5, -3, -1 and 0 dBm were used. The sensors transmitted one packet every second. At each power, the sensors transmitted 50 packets for statistical analysis. Upon data reception, the base station measured RSSI and LQI. The number of received packets was counted in order to compute PRR.

4.2.3 Experiments on location as a determination of necessary transmission power

The significance of the locations of the sending and receiving motes to determine the relationship between transmission power (TX) and reception quality is established. In this experiment, the base station location was the same whilst three sensors were placed at 10 different locations in the same direction with clear line-of-sight (LOS) including 1, 2, 3, 4, 5, 7, 10, 13, 16 and 20m. Each power adaptation cycle was ended after the maximum power had been reached. The other experimental parameters such as power levels, data sending rate and number of runs are stated in Section 4.2.2.

Fig. 2 shows the average RSSI readings of the three sensors at various locations and transmission power levels. The missing data indicate that the power provides RSSI reading less than -95dBm which is the minimum value reported by TinyOS. Fig. 3 shows average LQI readings of a sensor at various locations and transmission power levels. The missing data indicate unsuccessful data delivery.

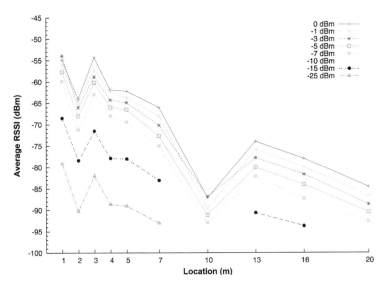

Fig. 2. Effects of sensor locations on RSSI

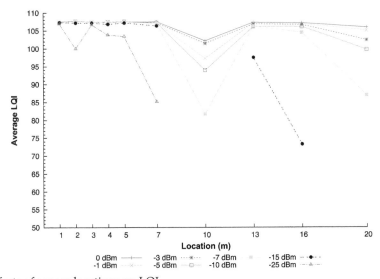

Fig. 3. Effects of sensor locations on LQI

According to Fig. 2 and Fig. 3, most of the RSSI measurements proportionally increased with the transmission power levels. Unlike the RSSI, the LQI measurements were stable at closer locations especially when higher power was used for transmission. Most of the LQI values decreased at greater distances. The minimum power level of -25dBm could be used to successfully deliver data to the base station only when the locations were within 7m. The decrease in received signal strength with increasing distances assumed in the prediction models do not apply in the results. For example, in the case of 2m, the sensor provides a weaker strength compared to a distance of 3m. The experimental results given in (Lin et al., 2006) and (Stoyanova et al., 2007) demonstrate similar observations on location effects. The RSSI and LQI are measured only when the base station receives data. The observed minimum RSSI values higher than -95 dBm indicate data reception.

4.2.4 Fluctuation in link quality metrics over time of day

This section investigates on how RSSI, LQI and PRR fluctuate over the time of day. The same base station and Sensor 1 were used. The sensor was located at 20m in the same environment. It transmitted one packet every second at 0 dBm for 1,440 minutes or 24 hours. The experiment was started in the morning before the office hour.

Fig. 4 demonstrates fluctuation of the RSSI, LQI and PRR over time of day. The RSSI fluctuated during the first half of the experiment. It was stable during the night time and the fluctuation was back later in the experiment. Unlike the RSSI, the LQI fluctuated throughout the experiment. At the beginning the PRR siginificantly decreased. This observation was resulted from the presence of people around the lab. The PRR increased during the night time as there were no staff and student in the lab.

In summary, apart from transmission power, location and heterogeneity in the manufacture, the link quality metrics are affected by the time-of-day. The presence of people around the lab is the main factor in this experiment and is considered as temporary physical barrier. Radio communication in WSN requires a line-of-sight. Some packets may be lost if there are some people in the sending path.

4.2.5 Relationship between metrics

This section aims to describe the relationships between RSSI, LQI and PRR. During packet reception, the base station measures RSSI and LQI. Apart from RSSI and LQI, the standard message type of TinyOS includes the CRC field which is a Boolean data type. The base station also looks at the CRC field to see if the data packet is received correctly. The numbers of data transmissions and receptions are counted to compute the PRR. This scheme can be used in a long-term operation.

However, the PRR may be estimated from the RSSI or LQI measurements. This concept suits a short term operation. The base station does not count the numbers of sent and received packets. Hence, the relationship between metrics needs to be established. Fig. 5 shows relationships between the link quality metrics at 5m, 12m and 19m. The average RSSI and LQI are computed at each transmission power level. The number of received packets is counted in order to calculate the PRR.

According to Fig. 5, several observations can be made as follows:

1. The PRR steeply increases with RSSI up to a certain point followed by more stable reliability measurements. Significant variations in reception rates are found when the RSSI readings are between -95 and -90 dBm. At least 95% PRR may be achieved at all distances if the sensor transmits data at the power producing RSSI greater than -90 dBm.

2. The higher LQI results in a more stable PRR. The relationship between LQI and PRR shown in Fig. 5 (b) is less clear than Fig. 5 (a). Similar results are also addressed in (Lin et al., 2006). According to these observations, RSSI should be used to relate to the PRR.

3. The LQI significantly increases with the RSSI. Convergence to particular LQI values is then observed. A lower bit error rate is observed when the base station receives packets with higher RSSI measurements.

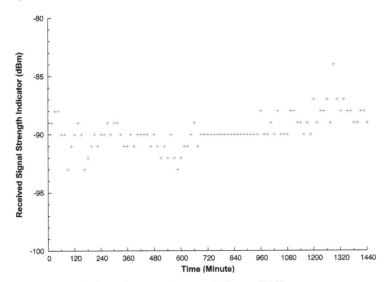

Fig. 4. (a) Fluctuation in link quality metrics over 24 hours RSSI

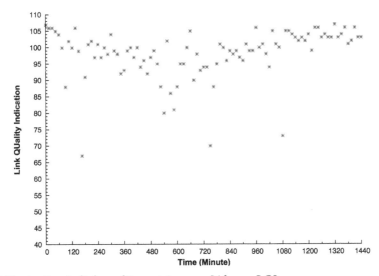

Fig. 4. (b) Fluctuation in link quality metrics over 24 hours LQI

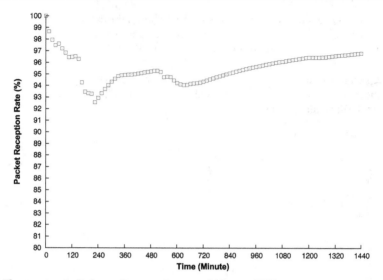

Fig. 4. (c) Fluctuation in link quality metrics over 24 hours PRR

The relationship between link quality metrics can be used to estimate an observed reliability from the measured receiving strength. This observation is addressed in (Lin et al., 2006) and (Srinivasan et al., 2006). After measuring the metrics, the base station determines whether the current transmission power requires an adaptation. The PRR steeply increases with the RSSI followed by significantly more stable measurements. The PRR should not be estimated from the RSSI between -95 to -90dBm as transmission power adaptation based upon this region will not be accurate. The measurements demonstrate that the network should operate at levels taken from an appropriate region.

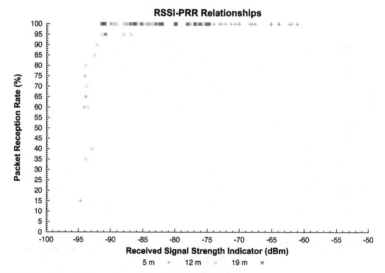

Fig. 5. (a) Relationships between metrics RSSI-PRR

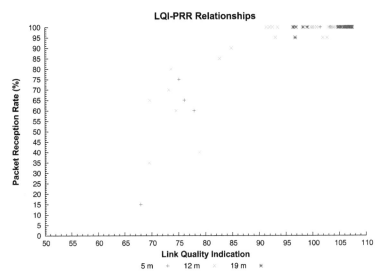

Fig. 5. (b) Relationships between metrics LQI-PRR

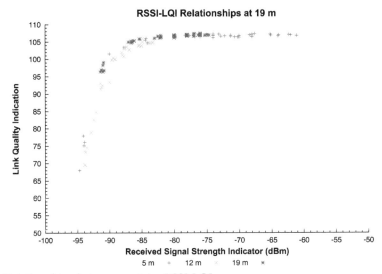

Fig. 5. (c) Relationships between metrics RSSI-LQI

4.3 Delays in wireless sensor network

This section provides some experimental results on delays in wireless sensor network (WSN) which affects PoRAP architecture development. Communication is represented by a frame structure which consists of several slots. A slot is assigned to each source and it transmits data when the allocated slot arrives. The slot length should be long enough to avoid data collisions at the base station where two packets from two different sources arrive approximately at the same time. Several experiments have been conducted in order to investigate some factors which affect the delays, including heterogeneity in sensor manufacturing and payload sizes.

4.3.1 Timestamp measurements and delay calculations

Details of timestamping scenario and delay calculations are given. As the base station does not know when the source is booted, at the beginning it broadcasts the control packet periodically. The periodic broadcast was set to 1 second. After the source is booted, it starts its transmission after the packet has been received. Similarly, the base station starts the next transmission after it has received the packet back from the source. Packet timestamping mechanisms and delay calculations are respectively illustrated in Fig. 6 and Table 4.

According to Fig. 6, the base station is booted at x_0. When the base station is ready to send, the timer is set to be fired at x_1 and send command is called at x_2. A timer is used in order to trigger packet transmission. Prior to transmission, the base station sets some fields in the message structure such as its id and transmission power. The SFD (Start of Frame Delimiter) transmission occurs at x_3. The timestamp is created and the packet payload content is modified to include the time of the transmission. Therefore, the fire-to-send and send command delays of the base station are equal to $x_2 - x_1$ and $x_3 - x_2$. The packet is completely transmitted by the radio at x_4 and the transmission delay is $x_4 - x_3$.

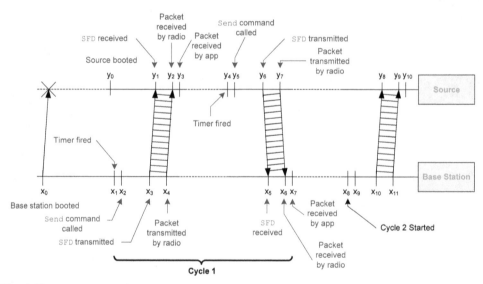

Fig. 6. Timestamp at various events

After being booted at y_0, the source receives the SFD at y_1. The receive event of the radio and application are signalled at y_2 and y_3 when the source receives the packet. The reception and receive delays of the base station are therefore $y_2 - y_1$ and $y_3 - y_2$. Once the packet has been received, the source requires some duration to process the information obtained from the packet. It then sets up its own transmission and the bits of packet are loaded into the radio buffer. The timer is fired at y_4 and the send command is called at y_5. The SFD is transmitted at y_6. Hence, the send command delay of the source is equal to $y_6 - y_5$. The transmission delay is $y_7 - y_6$. Table 4 summarises the delay calculations.

Delays	Calculations
Base Station	
• Fire-to-Send	$x_2 - x_1$
• Send Command Delay	$x_3 - x_2$
• Transmission	$x_4 - x_3$
• Reception	$x_6 - x_5$
• Receive	$x_7 - x_6$
Source	
• Reception	$y_2 - y_1$
• Receive	$y_3 - y_2$
• Fire-to-Send	$y_5 - y_4$
• Send Command Delay	$y_6 - y_5$
• Transmission	$y_7 - y_6$
Two-Way Propagation	$(x_5 - x_3) - (y_6 - y_1)$

Table 4. Summary of delay calculations

According to Table 4, the transmission and reception delays are calculated based upon when the events take place. The transmission delay is defined as the duration required for the radio to transmit the packet. In TinyOS 2.x, the CC2420Transmit interface provides a sendDone() event which notifies packet transmission completion. The reception delay is the duration required for packet reception by the radio, and the receive event is used for the timestamp. The fire-to-send delay indicates the desired interval for starting packet transmission after the timer is fired.

One Tmote Sky base station and one Tmote Invent source were used. The source was located at 0.5 m away from the base station. The base station was plugged into a desktop computer. In total 1,000 cycles of message exchange were run for each source. After the packet had been received, the node waited for 128ms and initiates its data transmission.

4.3.2 Experimental results

In order to consider the effects of payload size, an additional experiment was conducted. The scenario shown in Fig. 6 was used. All settings are the same except the payload sizes. In total five payload sizes were used including 39, 55, 75, 95 and 115 bytes. Note that the maximum payload for the CC2420 radio is limited to 117 bytes whilst the header size is 11 bytes. Send command and transmission delays of the source were determined. Two-way propagation delays were also computed. In the case of 39 bytes, reception and receive delays of source and base station were observed whilst all delays were observed for the larger payload sizes.

Statistical analysis of fire-to-send, send and transmission delays in milliseconds were conducted. The relationships between the 50th percentiles or medians of all sending delays and payload sizes are shown in Fig. 7. Note that "Send Command" delay is represented as "Send" in the figure. The results show that all delays increase with increasing payload sizes. The source requires more time to deliver larger packets to the radio. Similarly, larger packets require a longer duration for transmission. Increases in send command and transmission delays are greater than those of fire-to-send delay.

Statistical analyses of reception and receive delays in milliseconds were also made. The relationships between the 50th percentiles or medians of both receiving delays and payload sizes are shown in Fig. 8. Linear relationship between reception delay and payload size is also observed in Fig. 8. The receive delays are constant for all payload sizes.

The 32-KHz clock has been used in this experimental study and provides 32,768 ticks per second. There are 32 ticks in one millisecond. Therefore, the finest precision is approximately 0.03125 millisecond or 31.25 microseconds. The two-way propagation delays for all payload sizes are calculated and frequencies of the delay occurrences in ticks are shown in Table 5.

According to Table 5, frequencies of the 0-tick decrease with increasing payload sizes. Larger packets require more time to travel from source to destination. However, the two-way propagation delays are significantly less than the other delays.

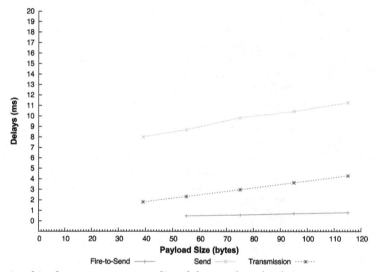

Fig. 7. Relationships between source sending delays and payload sizes

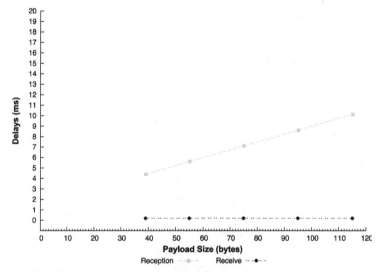

Fig. 8. Relationships between source receiving delays and payload sizes

Attribute		Payload Size (bytes)				
		39	55	75	95	115
Frequencies	0	858	807	785	755	740
	1	141	193	212	245	259
	2	0	0	3	0	0
Cycles		999	1,000	1,000	1,000	999

Table 5. Frequencies of two-way propagation delays

5. Design of PoRAP

This section describes the design of PoRAP (Power & Reliability Aware Protocol) which aims at minimising communication energy in wireless sensor network (WSN). The experimental results stated in previous section inform the design.

5.1 PoRAP main capabilities

In PoRAP, power can be conserved via transmission power adaptation and efficient medium access management. The selected link quality index is Received Signal Strength Indicator (RSSI) and it is measured by the base station during data reception. Along with the awareness of data loss, the adjusted power will often maintain the network operating at the region where data loss is minimised.

Additional communication can be saved by adopting the schedule-based MAC approach. Sending and receiving delays can be estimated as they are dependent upon packet size whilst two-way propagation delay is significantly small. Data transmissions are scheduled and the sources are mostly in sleep mode to conserve energy. Only one source engages the shared medium at a time for data transmission. Thus, data collision can be avoided and idle listening can be minimised. More explanations on PoRAP key capabilities are given as follows:

5.1.1 Schedule-based protocol

In the single-hop networks, sources are capable of communicating with their base station directly. This scenario is feasible when the sources and base station are located within communication range of each other. The base station may be connected to several sensors which require an access to the shared medium. Uncontrolled medium access possibly leads to data collisions at the base station. Collision is one of the main sources of power wastage in the WSN shared medium system. The medium access control (MAC) approach attempts collision avoidance. There are currently two main approaches proposed for WSN. Firstly, the medium is sensed to detect any ongoing activities in the medium before conducting data transmission and reception. This scheme is named contention-based.

PoRAP employs another approach in which each node is assigned a specific duration to use the shared medium. This scheme is called schedule-based. The other sensors cannot access and use the medium whilst a sensor is communicating within its time slot. Sources listen to the base station only once in a frame. Idle listening is therefore minimised. Moreover, data collisions at the base station can be avoided as there is only one source sending at a time. The slot length should be long enough to let the source and base station complete data transmission and reception. This scheme may not be suitable in the case of multi-hop WSN where each resource-constrained sensor has to maintain slot information of its neighbours. Furthermore, time synchronisation is required as both sender and receiver have to orchestrate the data communications to avoid collision caused by the other receivers.

Centralised scheduling control by the base station is feasible in PoRAP. Slot arrangement information can be sent to all sensors located in the range. The base station broadcasts a packet to all sources located in its range. Slot information such as number of slots, slot length and start time of first slot are included in the payload. Once the first frame is finished, the base station broadcasts again with the transmission power adaptation notification.

5.1.2 Communication power conservation

Power constraint should be taken into account when designing a protocol for WSN. Sensors may be left unattended after being deployed in the remote or hostile environment where battery recharge or replacement may be costly or infeasible. Communication accounts for power consumption in WSN. Several sensor platforms provide adaptation to the transmitting power and the concept of Transmission Power Control (TPC) has been adapted to WSN. The CC2420 radio employed by Tmote platform, which is used in this research, supports transmission power (TX) setting. The TX levels are stated by a 5-bit number. There are therefore 32 possible TX settings provided by the CC2420. In TinyOS, the setPower() command provided by CC2420Packet interface accepts a value between 0 to 31 for TX setting. However, the CC2420 datasheet specifies programmable TX in 8 steps from approximately -25 to 0dBm which are respectively equivalent to the power levels of 3 and 31. The Tmote datasheet follows guidelines given by the CC2420.

Transmission power adaptation policies in WSN should take application specifics into account. Different applications may require the sources to transmit data at different rates. For example, an environmental monitoring system may require the current temperature hourly whilst a surveillance system may require the data every second when an intrusion is detected. The sensors should be switched to sleep mode after transmission in order to minimise the idle listening. In a multi-hop network, each node is responsible for routing. It has to communicate with its neighbours to discover the best path by means of the least power utilisation. An amount of power is therefore required for listening in the multi-hop. However, a sensor in the single-hop scenario is capable of transmitting data directly to the base station. It may be switched to sleep mode after transmission. However, the source has to listen during the control slot transmission from the base station.

The power adaptation mechanisms in PoRAP do not require historic entries of RSSI and associated transmission power. The main reason is the limitation of buffering capacity of the radio chip. The base station should support a significant number of sources. In the CC2420 radio, the maximum buffer size is 128 bytes. Some bytes are required for the header and other controlling details. Only two bits are used to notify the power adaptation. The RSSI-PRR relationship obtained from the experimental studies is considered for adaptation as it suggests the operating region for WSN. In the case of power adaptation, the base station sets particular bits to notify the source. The sources get the bits and set their transmission power accordingly.

5.1.3 Link quality monitoring

Radio communication uses air as the transmission medium. There are several attributes ranging from differences in hardware components to environmental factors such as physical barriers which affect signal attenuation. Received signal strength estimation is unlikely as sensors can be placed in various areas of interest. An estimation model should not only determine distance between sender and receiver as an input, location should also be taken into account. A shorter distance may not always provide a higher received strength if a physical barrier appears in the communication line-of-sight (LOS). Moreover, the link quality metrics fluctuate over the time of day. The observed strength in an indoor environment may be lower during the nighttime. Applying the simple received signal strength estimation models, focusing mainly on distance and hardware properties, may not be sufficient. Therefore, PoRAP employs the measurement-based approach in order to more accurately adapt the transmission power.

Two link quality metrics are used in PoRAP. The RSSI is obtained by the radio chip whilst the PRR is specified by the applications. The relationship between RSSI and PRR can relate the application requirement to the observed link quality. As shown in Section 4.2.5, a clear relationship between the two metrics is established. The PRR steeply increases with the RSSI up to a certain point. The PRR is then stable after a certain value of RSSI and a lower RSSI or TX can be used to obtain the required PRR.

The range of required RSSI is obtained from the reliability requirement and the RSSI-PRR relationship. This range is recognised by the base station. Upon data reception, the base station measures the RSSI and compares it to the RSSI thresholds. The adaptation bits are set with respect to the comparison result. There are three available patterns of bit settings; the transmission power will be increased if the measured RSSI is lower than require and it will be decreased if the RSSI is higher. The sources will be notified to retain the current power if the RSSI is within the range.

5.2 PoRAP architecture

This section aims to describe PoRAP architecture. PoRAP aims at an efficient data delivery in WSN by means of energy conservation. Input of PoRAP comes from two external components, the user/application and the monitored phenomenon. PoRAP recognises the duty cycle and the awareness of data loss. The sensed data is another input and it will be sent from the source to the base station. In order to achieve the goals, the base station controls the sources whereas the sources send data to the base station. Required functionalities of the base station and the sources are then stated. The interactions between them are described and they are used to address the required components within the source and the base station. Moreover, the interactions between such components are also given in this section.

5.2.1 Overview of PoRAP

The main objective of PoRAP development is to provide an efficient data communication in WSN where the user/application has his/its own requirements such as reliability and duty cycle. The development of a generic network protocol for WSN is challenging as WSN are application specific. Fig. 9 shows an overview of PoRAP architecture in terms of the interactions between its main components.

According to Fig. 9, four main components are addressed including the user/application, sensed phenomenon, base station and sources. As WSN is application specific, the user/application has its own set of requirements. The base station directly interacts with the user/application whilst the sources collect physical directly from the phenomenon. The functionalities required at the base station and source can be listed as follows:

PoRAP

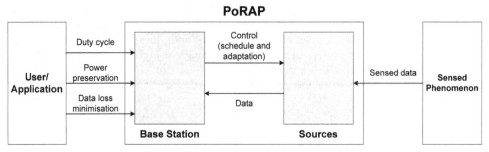

Fig. 9. Overview of PoRAP

Base station:

- **Recognise the requirements of user/application:** PoRAP aims at the low duty cycle application where the key objective is power conservation instead of throughput. Examples of this application category are habitat and environmental monitoring systems.
- **Control the source's operation:** This work focuses on the single-hop network where direct communication between sources and base station is feasible. No routing is required at each source and its operation is controlled by the base station in two aspects. Firstly, the base station determines whether transmission power used by the source needs to be adjusted by looking at the RSSI. Secondly, the communication cycle of each source is scheduled in order to avoid data collision and minimise idle listening.

Source:

- **Collect physical data:** WSN has been deployed to collect physical data from the targeted environment such as temperature and humidity. This work looks at how an efficient data delivery can be achieved by using lower transmission power whilst data loss is minimised. The processes of data collection are outside the scope of this study.
- **Data transmission:** After receiving the control information, the source sets two parameters. Firstly, it synchronises the communication schedule. Thus it will know when to start the radio for control reception and data transmission. Secondly, the source adapts its transmission power level according to the included notification. Lower power can be used and a significant amount of transmission power can be conserved.

Several interactions between the source and base station are required to achieve the functional requirements and they are addressed in Fig. 10.

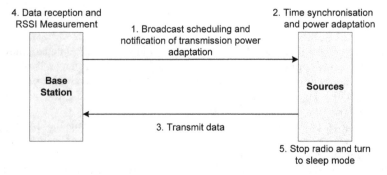

Fig. 10. Interaction between sources and base station

1. PoRAP focuses on the set of fixed sources which are located within communication range of the base station. The control packet includes scheduling and power adaptation notification and is broadcast to the sources using the maximum power level. This is feasible as the base station obtains extra power from the connecting computer.

2. Once the control packet is received by the source. Information on scheduling and notification is read. The source synchronises its schedule with the other nodes together with adjusting its transmission power accordingly.

3. After conducting time synchronisation and transmission power adaptation, the source waits for its slot to conduct data transmission using the adjusted transmission power. The radio must be started for communication.

4. The base station measures the RSSI during data reception. The observed RSSI is compared to the desired range which includes minimum and maximum values. The setting of the RSSI thresholds is obtained from the RSSI-PRR relationship. The selected RSSI should be obtained from the region where significant stability in the PRR is observed. The base station then decides whether transmission power adaptation is required. The notification is set accordingly.

5. The source stops its radio after transmission to save power. The amount of power consumption is the least when the source is in sleep mode. Timing is required for the source to start the radio again for the next communication cycle.

5.2.2 Components

The previous section points out several essential functions which are required to achieve the objectives of PoRAP development. This section aims to describe the essential components which give rise to this functionality. The selected operating system for WSN in this work is TinyOS which already provides several useful components and PoRAP takes those in TinyOS and adds some further modifications. The main components are determined from the interactions including the user/application, the observed phenomenon, the base station and source. Several components required at the base station and source are then considered. Moreover, the interactions between each component are demonstrated.

A) Components at base station and sources

The base station recognises the requirements of the user/application and controls the sources based upon the requirements. As PoRAP aims at the direct communication, the control information is broadcast to the sources which are located within the communication range. After physical data collection, the sources set their communication parameters prior to data transmissions. Fig. 11 depicts several components required at the base station and sources.

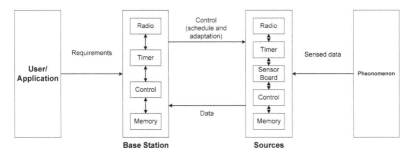

Fig. 11. Components at base station and sources

Each of the required components is described as follows:

- **Radio:** Each sensor employs the radio communication for wirelessly communicating with its neighbours or destinations. The radio has four major functions as follows:
  - o *Data communications:* Control information is sent by the base station's radio chip and is received by the source's radio chip. Data is sent by the source's radio chip and is received by the base station's radio chip.
  - o *Data buffering:* Prior to forwarding the received data to the higher layers or transmitting the data through the medium, the data is buffered. The buffering capacity is limited and dependent upon the radio chip. The capacity is important to the design of packet structures. For example, the control packet must not be longer than the allowable capacity but it has to carry all the required information.
  - o *Received signal strength measurement:* The received signal strength is important as it can reflect the current link quality. The latest radio chip provides the measurement of received signal strength such as Received Signal Strength Indicator (RSSI) and Link Quality Indication (LQI). RSSI is used in this work as it can be obtained from several radio models and its relationship with the Packet Reception Rate (PRR) is clear.
  - o *Transmission power adaptation:* The RSSI changes with transmission power and several factors such as location, time-of-day and environment. One of the main features in PoRAP is transmission power adaptation. The key concept is adjusting the current transmission power to achieve the power conservation and data loss minimisation. The latest radio model supports programmable transmission power.
- **Timer:** WSN is considered a share-medium system as all nodes have to access the medium prior to transmission. PoRAP aims at single-hop WSN where direct communication between source and base station is feasible. The sources are not responsible for routing. Instead of applying the contention-based scenario, the transmissions are scheduled. A slot is allocated for each source so that it can send only when its slot arrives. Otherwise, the radio is stopped and the source is switched to sleep mode for minimum energy consumption. A timer is therefore required for scheduling the radio start and stop.
- **Control:** It is used to control the other components especially when there is no control mechanism provided for some components. For example, an additional control interface is required for the radio and the interface is used to start and stop the radio.
- **Memory:** This component is the basic one which is also included in the sensor. Several variables along with their values and measurements are stored in the memory. For example, the required RSSI range which is obtained from the RSSI-PRR relationship. This range is stored in the memory and will be compared to the observed RSSI to determine whether any transmission power adaptation is required.
- **Sensor board:** This component is crucial for the sensors as it is responsible for collecting the physical data from the environment. The sensor board consists of several sensors such as temperature and humidity.

B) Interactions between components

This section aims at addressing the interactions between the components, and they are described in Fig. 12. The interactions within the base station and source can be separately described as follows:

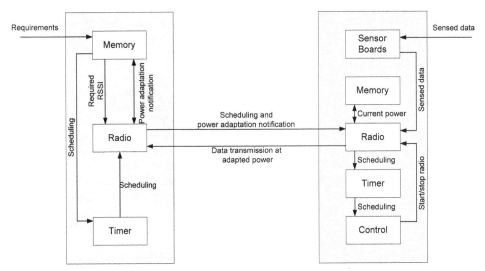

Fig. 12. Interactions between components

Base station

The base station acts as a destination for the data. The requirements are stored in the memory and they are used to set required RSSI range and the data sending rate. In PoRAP, the schedule-based scheme is adopted where each source has its own slot for data transmission. The slot must be large enough to accommodate several communication delays. According to the results in Section 4.3.2, sending and receiving delays are mainly dependent upon the packet size whereas the two-way propagation delay is significantly small. Models are required for estimating the slot size and they will be described later in this chapter. The next transmission begins after the other sources have already transmitted. Hence, PoRAP suits the applications which require a low duty cycle. The timer is used for scheduling the communications so it also uses this requirement from the application.

The required RSSI range can be obtained from the RSSI-PRR relationship which is dependent upon different conditions such as time-of-day, environment and location of deployment. The PRR is also used as an additional link quality metric as it is close to the reliability requirement. The main objective of PoRAP is to conserve communication energy whilst data loss is minimised. In the short term, the base station measures the RSSI when it receives the data packet. It uses the observed RSSI to determine whether power adaptation is required. The notification bits which are reserved for each source are then set. In the medium or longer term, the base station measures the PRR and uses that to determine what the upper and lower RSSI bounds should be. If more packets are lost, the RSSI bounds are increased. However, the bounds are slowly lowered to reduce power expenditure if the loss is low or non-existence. The number of notification bits is crucial as the base station has to communicate with all the sources in its range. Using too many bits may lead to a control packet which is larger than the buffering capacity of the radio chip.

The base station radio is not started or stopped as it has to continually receive the data packets from its sources. Data packet receptions occur after broadcasting the control packet at the maximum transmission power level. This concept is feasible as the base station has an extra source of power from its connecting computer. In PoRAP, the power conservation goal is mainly located at the sources.

Source

In WSN, the source is responsible for physical data collection. The data is then transmitted to the base station. The key objective of PoRAP is to conserve communication power of the source. Prior to transmission, the source determines whether it has to adapt its current power. The notification is included in the control packet and it is received by the radio of the source. As the buffering capacity of the radio is limited, the base station notifies what the source should do to its current power instead of specifying the appropriate power level. Thus, the source has to store the current power in the memory. For example, the current power is increased if a lower RSSI is measured by the base station. Moreover, the source should recognise the limitations of the transmission power adaptation. The base station may need its source to increase the power even if the maximum has already been reached. The minimum and maximum power levels are dependent upon the selected radio chip.

Apart from the power adaptation signaling, the scheduling is also included in the control packet. Time synchronisation is crucial in the schedule-based approach. The local clock of each node may run at different speeds. In PoRAP, the sources synchronise with their base station. The synchronisation refers to several timestamps which are conducted at the MAC layer where hardware and operating system dependent delays can be disregarded. The scheduling is also recognised by timer and controls components. Several timers are required as they are responsible for timing the sending and receiving communications. The timers operate closely with the control in order to start and stop the radio. For example, the radio is stopped after the data packet is sent. The source knows when it has to wake up to receive the next control packet. The timer is then started, counting the generated ticks. A control interface is used to start the radio for control reception when the scheduled time has come.

5.2.3 Transmission power adaptation policies

A sensor consists of hardware components working together to facilitate sensing, processing and communicating tasks. Amongst these components, the transceiver or radio unit is responsible for data communication. Normally, the radio unit supports programmable transmission power and the possible adaptable range is given in the datasheet. For example, the Tmote sensor platform which is chosen for this work employs the CC2420 radio. The minimum and maximum powers are 0 and -25dBm, respectively. There are two main factors which should be taken into account when transmission power adaptation is required. Several hardware limitations of the radio unit include the allowable minimum, maximum transmission power and base noise. The environmental factors leading to signal strength attenuation should be determined. The selected transmission power should be high enough to produce the associated receiving strength which is not discarded by the receiving node. The maximum power allowed by the radio unit is used as the upper limit. In PoRAP, sources use maximum power for their first transmissions. This policy ensures that the packets will likely be transmitted to the base station. However, both base noise and attenuation are respectively hardware and environment dependent. It is difficult to specify an accurate power adaptation level which can be generally used. Moreover, additional resources will be required if the sources periodically measure and send their base noise to the base station. Attenuation is hard to predict as link quality changes over time. Hence,

PoRAP repetitively increases or decreases the transmission power within an allowable range instead of discovering the right power.

5.2.4 Frame structure and slot decomposition

In PoRAP, a frame is used to represent a communication cycle which consists of one control slot at the beginning followed by several data slots. Its structure is shown in Fig. 13.

G indicates the guard of the frame and is used to protect frame overlapping. A control slot is used by the base station for broadcasting control data which includes scheduling information and transmission power (TX) adaptation notification to its sources. The slot information is required by the sources in order to synchronise themselves to the base station. The time of starting the first data slot is required so that the sources know when data is sent. In PoRAP, each slot has the same length which should accommodate a specific data payload size to be completely transmitted and received.

Fig. 13. Frame structure

According to Fig. 13, the sources firstly turn their radios on during the control slot to receive the control information. If they are not assigned to the first data slot, they stop the radios after knowing when their slots start. When their slots arrive, the radios are re-started to send the data. Unlike sources, the base station listens to the medium for data packet reception all the time. The decomposition of a slot is depicted in Fig. 14.

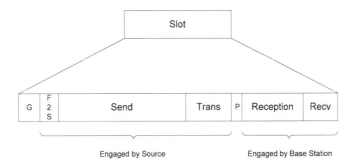

Fig. 14. Data slot decomposition

There are four main delay components in Fig. 14. The G and P are respectively the guard time and propagation delay. The first component is the guard length which prevents the slots from overlapping. Feasible overlapping scenarios together with guard time consideration are provided later in this section. The second component consists of fire-to-send (F2S), send and transmission delays and this is the sending delay component. This component is caused by the source. The third one is propagation delay which is considerably smaller than the other delays. Finally, the receiving delay component includes the reception and receive delays. This component is considered during packet arrival at the base station.

5.2.5 Estimation of communication delays

A schedule-based approach is adopted in PoRAP. The base station allocates and manages several time slots. In this work, a set of fixed nodes is determined. The number of data slots is therefore equal to the number of booted sources which are able to receive the control packet broadcast by the base station. The source initiates transmission when its assigned slot arrives. Apart from data slots, a frame also contains a control slot which is used by the base station. The slot must be large enough to accommodate sending and receiving delays to avoid feasible data collisions. As shown in Section 4.3.2, the delays are dependent upon packet sizes. This section analyses these relationships for delay estimations.

The experimental results on delays described in Section 4.3.2 demonstrate linear relationships between delays and data packet sizes. The key objective in this part is to discover the two coefficients obtained from linear regression analyses. The coefficients will be used to establish the models providing estimated delays where payload sizes are input. In total 5 payload sizes including 39, 55, 75, 95 and 115 bytes were varied to investigate changes in delays. Regression analyses have been applied to the results of the sending and receiving delays of the source and the base station. As linear relationships between delays and payload sizes are observed, two coefficients of the linear equation (c_0 and c_1) are the required output where c_0 is the y-intercept and c_1 is the slope. The Table 6 summarises the coefficients of each delay.

According to Table 6, the coefficients for the base station do not significantly differ from those for the source. The fire-to-send delays of the base station are constant whilst the source provided a linear relationship.

Delays	Measured at	Coefficients	
		c_0	c_1
1. Fire-to-send (F2S)	Base station	Constant delays of 0.50 ms	
	Source	0.204	0.025
2. Send	Base station	11.367	0.043
	Source	11.263	0.043
3. Transmission	Base station	0.490	0.033
	Source	0.552	0.033
4. Reception	Base station	1.521	0.076
	Source	1.521	0.076
5. Receive	Base station	Constant delays of 0.22 ms	
	Source	Constant delays of 0.22 ms	

Table 6. Coefficients obtained from experimental results at 99th percentile

In the case where the payload size is zero, a specific duration is still required for header transmission and reception. For CC2420, the header is approximately 11 bytes and requires 0.352ms for the delivery. An additional duration is required for transmitting processes which can be considered as an overhead. The send delay is the largest of the experimental results. It is an interval from calling the send() command until capturing the SFD. Several mechanisms undertaken by the application software and operating system to facilitate the sending also require time and are included in the send delay. For example, when the send() command is called by the application, an interrupt is signaled to TinyOS. The packet is buffered and the CC2420 is switched to transmitting mode. This sending overhead due to software manipulation and hardware setup is regardless of payload size. Increases in payload size require additional delays. For example, for every byte increase in the payload size, the send and reception delays of a source respectively increase by 0.043 and 0.076ms. However, the payload size does not affect receive delay. The coefficients can be used to estimate the communication delays.

5.2.6 PoRAP scenario

PoRAP is developed to effectively support data communication in single-hop wireless sensor network (WSN). The base station communicates with its sources for controlling and data collection purposes. As the base station does not know when each source is booted, a setup process is required at the beginning of frame structure. Acting as a data receiver, the base station always listens to the medium for incoming messages after broadcasting the control packet. Hence, the base station desires extra power which can be obtained from external sources such as a desktop or laptop computer.

A) Control and setup phase

Prior to data transmission, the sources have to setup their parameters based upon the control information received from their base station. The information such as number of slots, slot length and slot start time is used to control the sources in order to send data within an allocated slot at an adapted transmission power. As the base station has no information on when the sources join the network, it has to discover which sources are booted and ready for communication. In the control and setup phase, the base station periodically broadcasts control packets to all sources located in its communication range. The broadcasted packet is received by the booted sources and they use the received information to setup the communication parameters.

There are three main parts to the control information included in the control packet. The first attribute indicates the identification of the base station. This field supports a future enhancement of PoRAP which supports the multiple base station system. It can be also used to differentiate between the control and data packets. The second attribute is schedule related. Some information is required by the sources in order to synchronise with their base station. These parameters include the number of slots, slot length and the start time of the first slot. The base station specifies the slot start time with respect to the Start of Frame Delimiter (SFD) transmission in order to minimise the effects of application and hardware processing delays. The source assigned to the first data slot sets its timer to fire and sends data when the time arrives. Other sources start at different times and they compute the starting times from the slot information. The transmission parameters are required to be completely set before the phase begins. Slot length determination for data slot can therefore be applied to the control slot. The base station periodically broadcasts its control packet. There are two main objectives of periodic broadcasting are maintaining synchronisation between nodes and supporting changes in network topology. Additional sources may be booted during the frame and some sources may be running out of energy. The number of sources is therefore modified by the base station.

B) Data delivery phase

Slots are allocated by the base station in order to facilitate data transmissions of the sources. The data delivery phase starts after the control packet is received by the sources. The number of slots is fixed as it assumes that the base station communicates with the fixed number of sources and the number is constant throughout the operation. Data collected by the sources is stored in the data packet and is delivered to the base station.

The Received Signal Strength Indicator (RSSI) is measured when the base station receives the data. The RSSI linearly relates to the transmission power and the RSSI-PRR relationship is established in Fig. 5 (a). The PRR steeply increases with the RSSI up to a certain point. The increase in PRR then becomes insignificant or it becomes constant after this point. The RSSI is monitored and compared to the desired range. Power adaptation notification is conducted by the base station. The sources are notified by control packet reception in the next frame.

Apart from data, the identification (id) of a source is also included in the data packet. Specifying source id is an important issue and it may be done in several ways. For example, the SFD of the control packet reception time may be modified to obtain the id. However, sensors are considered resource constrained. Simple calculations should be included in the sources. Within the 128-byte buffering limitation in CC2420, one to two bytes should be enough to represent the id. Furthermore, the id can be assigned at installation time. Prior to deployment, a particular id is allocated to the source. For example, an id of 1 may be used for installing PoRAP in the first source in the network. Additional power conservation is introduced during the data delivery phase. The strategy benefits from adopting the time-slot based concept. As sources know when to receive control and to transmit data packets, it is possible to periodically turn the radio on for such periods. Fig. 15 describes the mode switching concept during the data delivery phase. The C&S, R, S and G represent control and setup, receive, send and guard, respectively.

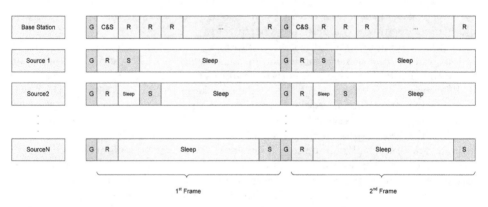

Fig. 15. Mode switching during the data delivery phase

According to Fig. 15, each source is in wakeup mode when its radio is turned on for two reasons; control packet reception and data packet transmission. Otherwise, its radio is turned off and the source is switched to sleep mode. However, the base station radio is always turned on. This strategy minimises idle listening power at the sources.

6. PoRAP energy conservation evaluation

An experiment was conducted in a 16m x 20m indoor environment to evaluate the energy conservation of PoRAP. A network consisting of 20 sources and a base station was set up. Tmote Sky motes were used as both sources and base station. The sources were placed at 20 different locations with 14 different distances and the base station was connected to a desktop machine. All motes had the same height above ground level and had the same antenna orientation. The minimum and maximum distances are 1 and 22.5m, respectively.

Initially, the base station broadcast its 18-byte control packet to the sources. The sources then transmitted the 48-byte data packets back to the base station. A communication cycle was completed after the base station had received the data from all sources. Apart from the maximum power settings, four additional RSSI settings are included. The minimum RSSI thresholds were set to -90, -80, -70 and -60dBm whereas the corresponding maximum thresholds were -80, -70, -60 and -50dBm, respectively. The power is not adapted if the measured RSSI is between the thresholds and the aim is to obtain nearly 100% PRR. Each mote transmitted every 5 minutes and the experiment lasted for 24 hours. The results are shown in Table 7.

Dist. (m)	-90 < RSSI < -80		-80 < RSSI < -70		-70 < RSSI < -60		-60 < RSSI < -50		Max TX	
	Saved Trans Current	Packet Loss (%)	Saved Trans Current	Packet Loss (%)	Saved Trans Current	Packet Loss (%)	Saved Trans Current	Packet Loss (%)	Saved Trans Current	Packet Loss (%)
1	51.2	0	51.2	0	51.2	0	35.6	0	0	0
2	51.2	0.3	35.6	0.7	0	0	0	0	0	0
4	43.1	2.3	43.1	0.7	28.2	0	0	0	0	0
6	43.1	4.7	28.2	0	0	0.3	0	0	0	0
8	51.2	5	0	0.7	0	0.3	0	0	0	0
10	51.2	5.3	35.6	0	0	0	0	0	0	0
12	51.2	5.7	20.1	0	0	0	0	0	0	0
14	0	14	28.2	0	0	0	0	0.4	0	3.7
16	28.2	5.7	20.1	0	0	0	0	0	0	1.2
20	43.1	3.7	0	0.7	0	0	0	1.2	0	2.1

Table 7. Conserved transmitting current and data packet loss

According to Table 7, lower RSSI settings result in higher percentage of packet loss and conserved transmitting power. Lower power is used to produce the required RSSI range. A significant amount of power up to 50% can be yielded. However, the highest packet loss is obtained when the RSSI is between -90 and -80 dBm.

7. Conclusion

This chapter describes several aspects which should be considered during developing a network protocol for wireless sensor network (WSN). WSN has been used in both surveillance and civil applications. It is considered application specific as each application has its own set of requirements. Two main categories are proposed including event-based and periodic-based application. Throughput is the key requirement in the event-based whilst lifetime is the key in the periodic-based. Moreover, one of major drawbacks of WSN is resource constraint. The power for all operations comes from tiny batteries. Under some circumstances, it is uneconomical or impractical to change or recharge the batteries. In WSN, the data is delivered via wireless link which is susceptible to the surrounding environments. The radio unit is responsible for data delivery has a limited buffering capacity. Control information should be minimised to be included in a packet.

The Power & Reliability Aware Protocol (PoRAP) is developed and its main objective is to provide an efficient data communication by means of energy conservation whilst reliability is maintained. Its three key elements include direct communication, adaptive transmission power and intelligent scheduling. With adaptive transmission power and intelligent scheduling, the power consumption is minimised as a result of a lower transmitting power, collision avoidance and minimised idle listening without unnecessary data losses. The key capabilities of PoRAP make it suitable for use in the periodic-based WSN applications with regular reporting patterns where maximising bandwidth is not the prime concern. PoRAP thus applies to some of the WSN applications such as environmental and habitat monitoring where the sources often remain at their positions throughout the operation. Slots are allocated to the sources for data transmissions. In PoRAP, it is assumed that the number of allocated slots is equal to that of sources. A low duty cycle application is more efficient using PoRAP when the percentage of slot usage is high. The evaluation results indicate up to 50% of power can be yielded whilst the reliability is within the desired range. However, PoRAP is not applicable if a source has to wait longer until the next cycle is started. Therefore, a limitation of PoRAP arises when there is a high slot overhead because there are many sources in the network.

8. References

Warneke, B. & Pister, K.S.J. (2002). MEMS for Distributed Wireless Sensor Networks, *Proceeding of the 9th International Conference on Electronics, Circuits and Systems.* Dubrovnik, Croatia

Mainwaring, A.; Polasrte, J. ; Szewczyk, R.; Culler, D. & Anderson, J. (2002). Wireless Sensor Networks for Habitat Monitoring, *WSNA'02*, Atlanta, Georgia, USA.

Allen, G.W.; Lorincz, K.; Ruiz, A.; Marcillo, O.; Johnson, J.; Lees, J. & Welsh, M. (2006). Deploying a Wireless Sensor Network on an Active Volcano. *IEEE Internet Computing*, Vol.10, No.2, pp.18-25

Essa, I.A. (2000). Ubiquitous Sensing for Smart and Aware Environments. *IEEE Personal Communications*

Srivastava, M.; Muntz, R. & Potkonjak, M. (2001). Smart Kindergarten: Sensor-Based Wireless Networks for Smart Developmental Problem-Solving Environments. *ACM SIGMOBILE*, Rome, Italy

Jovanov, E.; O'Donnell Lords, A.; Raskovic, D.; Cox, P.G.; Adhami, R. & Andrasik, F. (2003). Stress Monitoring Using a Distributed Wireless Intelligent Sensor System. *IEEE Engineering in Medicine and Biology Medicine*

Otto, C.; Milenković, A.; Sanders, C. & Jovanov, E. (2006). System Architecture of a Wireless Body Area Sensor Network for Ubiquitous Health Monitoring, *Journal of Mobile Multimedia*, Vol.1, No.4, pp.307-326

Arora, A.; Dutta, P.; Bupat, S.; Kulathumani, V.; Zhang, H.; Naik, V.; Mittal, V.; Cao, H.; Demirbas, M.; Gouda, M.; Choi, Y.; Herman, T.; Kulkarni, S.; Arumugam, U.; Nesterenko, M.; Vora, A. & Miyashita, M. (2004). A Line in the Sand: A Wireless

Sensor Network for Target Detection, Classification, and Tracking. *Computer Networks: The International Journal of Computer and Telecommunications Networking*, Vol.46, Issue 5, pp.605-634

Chintalapudi, K.; Fu, T.; Paek, J.; Kothari, N.; Rangwala, S.; Caffrey, J.; Govindan, R.; Johnson, E. & Masri, S. (2006). Monitoring Civil Structures with a Wireless Sensor Network, *IEEE Internet Computing*

Juang, P.; Oki, H.; Wang, Y.; Martonosi, M.; Peh, L.S.; & Rubenstein, D. (2002). Energy-Efficient Computing for Wild-Life Tracking: Design Tradeoffs and Early Experiences with ZebraNet, *ASPLOS'02*, ACM

Szewczyk, R.; Mainwaring, A.; Polasrte, J.; Anderson, J. & Culler, D. (2004). An Analysis of a Large Scale Habitat Monitoring Application, *SenSys'04*, Baltimore, Maryland, USA

Martinez, K.; Padhy, P.; Riddoch, A.; Ong, H.L.R. & Hart, J.K. (2005). Glacial Environment Monitoring Using Sensor Networks, *REALWSN'05*, Stockholm, Sweden

Kottapalli, V.A.; Kiremidjian, A.S.; Lynch, J.P.; Carryer, E.; Kenny, T.W.; Law, K.H. & Lei, Y. (2003). Two-Tiered Wireless Sensor Network Architecture for Structural Health Monitoring, *Proceedings of SPIE's 10th Annual Symposium on Smart Structures and Materials*, San Diego, USA

Paek, J.; Chintalapudi, K.; Caffrey, J.; Govindan, R. & Masri, S. (2005). A Wireless Sensor Network for Structural Health Monitoring: Performance and Experience, *Proceedings of the 2nd IEEE Workshop on Embedded Networked Sensors (EmNetS-II)*, Sydney, Australia

Schmid, T.; Dubois-Ferrière, H. & Vetterli, M. (2005). SensorScope: Experiences with a Wireless Building Monitoring Sensor Network, *REALWSN'05*, Stockholm, Sweden

Dreicer, J.S.; Jorgensen, A.M. & Dors, E.E. (2002). Distributed Sensor Network with Collective Computation for Situational Awareness, *AIP Conference Proceedings*, Vol.632, pp.235-243

Simon, G.; Balogh, G.; Pap, G.; Maróti, M.; Kusy, B.; Sallai, J. ; Lédeczi, Á.; Nádas, A. & Frampton, K. (2004). Sensor Network-Based Countersniper System, *SenSys'04*, Maryland, USA

Coleri, S.; Cheung, S.Y. & Varaiya, P. (2004). Sensor Networks for Monitoring Traffic, *University of California Berkeley Technical Report*, August 2004

Brignone, C.; Conners, T.; Lyon, G. & Pradhan, S. (2005). SmartLOCUS: An Autonomous, Self-Assembly Sensor Network for Indoor Asset and Systems Management, *Hewlett-Packard Development Company Technical Report*, June 2005

Sankarasubramaniam, Y.; Akan, O.B. & Akyildiz, I.F. (2003). ESRT: Event-to-Sink Reliable Transport in Wireless Networks, *ACM MobiHoc'03*, Maryland, USA

Ee, C.T. & Bajcsy, R. (2004). Congestion Control and Fairness for May-to-One Routing in Sensor Networks, *ACM SenSys'04*, Bultimore, Maryland, USA

Hull, B.; Jamieson, K. & Balakrishnan, H. (2004). Mitigating Congestion in Wireless Sensor Networks, *ACM SenSys'04*, Bultimore, Maryland, USA

Lu, C.; Blum, B.M.; Abdelzaher, T.F.; Stankovic, J.A. & He, T. (2002). RAP: A Real-Time Communication Architecture for Large-Scale Wireless Sensor Networks, *RTAS*, September 2002

Wan, Chieh-Yih; Eisenman, S.B. & Campbell, A.T. (2003). CODA: Congestion Detection and Avoidance in Sensor Networks, *ACM SenSys'03*, Los Angeles, USA

Wan, Chieh-Yih; Campbell, A.T. & Krishnamurthy, L. (2002). PSFQ: A Reliable Transport Protocol for Wireless Sensor Networks, *ACM WSNA'02*, Atlanta, Georgia, USA

Stann, F. & Heidemann, J. (2003). RMST: Reliable Data Transport in Sensor Networks, *IEEE International Workshop on Sensor Net Protocols and Applications (SNPA)*, Anchorage, USA

Intanagonwiwat, C.; Govindan, R.; Estrin, D. & Heidemann, J. (2003). Directed Diffusion for Wireless Sensor Networking, *IEEE/ACM Transactions on Networking*, Vol.11, No.1, February 2003

Xu, N.; Rangwala, S.; Chintalapudi, K.K.; Ganesan, D.; Broad, A.; Govindan, R. & Estrin, D. (2004). A Wireless Sensor Network for Structural Monitoring, *ACM SenSys'04*, Baltimore, Maryland, USA

Polastre, J.; Hill, J. & Culler, D. (2004). Versatile Low Power Media Access for Wireless Sensor Networks, *SenSys'04*, November 2004

Tolle, G.; Polastre, J.; Szewczyk, R.; Culler, D.; Turner, N.; Tu, K.; Burgess, S.; Dawson, T.; Buonadonna, P.; Gay, D. & Hong, W. (2005). A Macroscope in the Redwoods, *SenSys'05*, November 2005

Shnayder, V.; Hempstead, M.; Chen, Bor-rong; Allen G.W.; & Welsh, M. (2004). Simulating the Power Consumption of Large-Scale Sensor Network Applications, *SenSys'04*, Baltimore, Maryland, USA

Lin, S.; Zhang, J.; Zhou, G.; Gu, L.; He, T. & Stankovic, J.A. (2006). ATPC: Adaptive Transmission Power Control for Wireless Sensor Networks, *SenSys'06*, Boulder, Colorado, USA

Stoyanova, T.; Kerasiotis, F.; Prayati, A. & Papadopoulos, G. (2007). Evaluation of Impact Factors on RSS Accuracy for Localization and Tracking Applications, *In MobiWac'07*, October 2007

Srinivasan, K.; Dutta, P.; Tavakoli, A. & Levis, P. (2006). Understanding the Causes of Packet Delivery Success and Failure in Dense Wireless Sensor Networks, *Technical Report SING-06-00*, Stanford University

7

Standardised Geo-Sensor Webs for Integrated Urban Air Quality Monitoring

Bernd Resch, Rex Britter, Christine Outram, Xiaoji Chen and Carlo Ratti

Massachusetts Institute of Technology,
USA

1. Introduction

'In the next century, planet earth will don an electronic skin. It will use the Internet as a scaffold to support and transmit its sensations. This skin is already being stitched together. It consists of millions of embedded electronic measuring devices: thermostats, pressure gauges, pollution detectors, cameras, microphones, glucose sensors, EKGs, electroencephalographs. These will probe and monitor cities and endangered species, the atmosphere, our ships, highways and fleets of trucks, our conversations, our bodies – even our dreams.' (Gross, 1999)

Following this comprehensive vision by Neil Gross (1999), it can be assumed that sensor network deployments will increase dramatically within the coming years, as pervasive sensing has recently become feasible and affordable. This enriches knowledge about our environment with previously uncharted real-time information layers.

However, leveraging sensor data in an ad-hoc fashion is not trivial as ubiquitous geo-sensor web applications comprise numerous technologies, such as sensors, communications, massive data manipulation and analysis, data fusion with mathematical modelling, the production of outputs on a variety of scales, the provision of information as both hard data and user-sensitive visualisation, together with appropriate delivery structures. Apart from this, requirements for geo-sensor webs are highly heterogeneous depending on the functional context.

This chapter addresses the nature of this supply chain; one overarching aspect is that all elements are currently undergoing both great performance enhancement combined with drastic price reduction (Paulsen & Riegger, 2006). This has led to the deployment of a number of geo-sensor networks. On the positive side the growing establishment of such networks will further decrease prices and improve component performance. This will particularly be so if the environmental regulatory structure moves from a mathematical modelling base to a more pervasive monitoring structure.

Of specific interest in this chapter is our concern that most sensor networks are being built up in monolithic and specific application-centred measurement systems. In consequence, there is a clear gap between sensor network research and mostly very heterogeneous end user requirements. Sensor network research is often dedicated to a long-term vision, which tells a compelling story about potential applications. On the contrary, the actual implementation is mostly not more than a very limited demonstration without taking into account well-known issues such as interoperability, sustainable development, portability or the coupling with established data analysis systems.

Therefore, the availability of geo-sensor networks is growing but still limited. Deborah Estrin pointed out in 2004 that no real sensor network applications exist, apart from short-lived prototypical and very domain-specific demos (Xu, 2004). Recently, some examples of urban geo-sensor networks arose, but the clear social, health and economic benefits have not been demonstrated or described in a manner that would compel this sort of investment in particular for urban environments.

The goal of the Common Scents project is that its highly flexible architecture will bring sensor network applications one step further towards the realisation of the vision of a 'digital skin for planet earth' and have particularly far-reaching impacts on urban monitoring systems through the deployment of ubiquitous and very fine-grained sensor networks. In other words, the broad goal of the project is to develop an overarching infrastructure for various kinds of sensor network applications.

After this brief introduction the chapter provides a discussion of related work. Thereafter, we present the Common Scents approach and implementation for urban monitoring applications. Then, we illustrate some possible application areas for the system. We treat challenges for the deployment of sensor networks that are specific to the city and analyse how such systems can change the city as the functional context by adding new unseen information layers. We conclude with a short summary, discussion and future outlook.

2. Related work

The first domain of related work is **sensor network** development for environmental monitoring. The Oklahoma City Micronet (University of Oklahoma, 2009) is a network of 40 automated environmental monitoring stations across the Oklahoma City metropolitan area. The network consists of 4 Oklahoma Mesonet stations and 36 sites mounted on traffic signals. At each traffic signal site, atmospheric conditions are measured and transmitted every minute to a central facility. The Oklahoma Climatological Survey receives the observations, verifies the quality of the data and provides the data to Oklahoma City Micronet partners and customers. One major shortcoming of the system is that it is a very specialised implementation not using open standards or aiming at portability. The same applies to CORIE (Center for Coastal and Land-Margin Research, 2009), which is a pilot environmental observation and forecasting system (EOFS) for the Columbia River. It integrates a sensor network, a data management system and advanced numerical models.

Paulsen (2008) presents a sensing infrastructure called PermaGIS that attempts to combine sensor systems and GIS-based visualisation technologies. The sensing devices, which measure rock temperature at ten minute intervals, focuses on optimising resource usage, including data aggregation, power consumption, and communication within the sensor network. In its current implementation, the infrastructure does not account for geospatial standards in sensor observations. The visualisation component uses a number of open standards (OGC WMS, WFS) and open-source services (UMN Map Server, Mapbender).

There are a number of approaches to **leveraging sensor information in GIS** applications. Kansal et al. (2007) present the SenseWeb project, which aims to establish a Wikipedia-like sensor platform. The project seeks to allow users to include their own sensors in the system and thus leverage the 'community effect', building a dense network of sensors by aggregating existing and newly deployed sensors within the SenseWeb application. Although the authors discuss data transformation issues, data fusion, and simple GIS analysis, the system architecture is not based on open (geospatial) standards, only standard web services. The web portal implementation, called SensorMap (Nath et al., 2006), uses the Sensor Description Markup Language (SDML), an application-specific dialect of the OGC SensorML standard.

A GIS mash-up for environmental data visualisation is presented in the nowCOAST application (National Oceanic and Atmospheric Administration, 2011). Data from providers such as National Oceanic and Atmospheric Administration (NOAA), U.S. Geological Survey (USGS), Department of Defence (DOD) or local airports, are integrated in a web-based graphical user interface. nowCOAST visualises several types of raw environmental parameters and also offers a 24-hour sea surface temperature interpolation plot with 0.5 degree spatial resolution, produced using a two-dimensional variation interpolation scheme. The system does not make use of open standards for sensor measurements and data provision.

The second related research area is real-time data integration and **sensor fusion** for geospatial analysis systems. Harrie (2004) and Lehto & Sarjakoski (2005) present web services based on the classic request/response model. Although both methods widely use open GIS standards, they are not suitable for the integration of real-time data for large volumes of data. Sarjakoski et al. (2004) establish a real-time spatial data infrastructure (SDI), which performs several application-specific steps, such as coordinate transformation, spatial data generalisation, query processing or map rendering and adaptation. However, the implemented system accounts neither for event-based push mechanisms nor for the integration of sensor data.

Other approaches for real-time data integration try to tackle the issue from a database perspective. Oracle's system, presented by Rittman (2008), is essentially a middleware between (web) services and a continuously updated database layer. Like Sybase Inc. (2008), the Oracle approach is able to detect database events in order to analyse heterogeneous data sources and trigger actions accordingly. Rahm et al. (2007) present a more dynamic method of data integration and fusion using on-the-fly object matching and metadata repositories to create a flexible data integration environment.

The third related research area is the development of **an open data integration system architecture** in a non-application specific infrastructure. Srivastava et al. (2006) and Balazinska et al. (2007) present general concepts in a systems architecture and data integration approach but there are no concrete conclusions as to how the final goal of establishing such an infrastructure could be achieved. A more technical approach for ad-hoc sensor networks is described by Riva & Borcea (2007), where the authors discuss challenges to making heterogeneous sensor measurements combinable through the creation of highly flexible middleware components. The method is application-motivated and thus very detailed as far as specific implementation details are concerned.

3. Common Scents measurement infrastructure

The *Common Scents* project aims at developing an interoperable open standards based infrastructure for providing fine-grained air quality data to allow users to assess environmental conditions instantaneously and intuitively. The goal is to provide citizens with unseen up-to-date information layers in order to support their short-term decisions. To achieve this vision, we utilise the CitySense sensor testbed and technologically build our system upon the *Live Geography* approach presented by Resch et al. (2009a). It should be stated that the term 'real-time' in our case is not defined by a pre-set numerical time constant, but more by qualitative expressions such as 'immediately' or 'ad-hoc', i.e. information layers have to be created in a timely manner to serve application-specific purposes. (Resch et al., 2009b)

3.1 Design principles

In the conception of our technical infrastructure we accounted for different design principles (Service Oriented Architectures – SOA, modular software infrastructures etc.) to ensure flexibility, reusability and portability of the components and the overall infrastructure. Figure 1 shows the modular architecture and the standardised service interfaces that are used to connect the different components in the workflow.

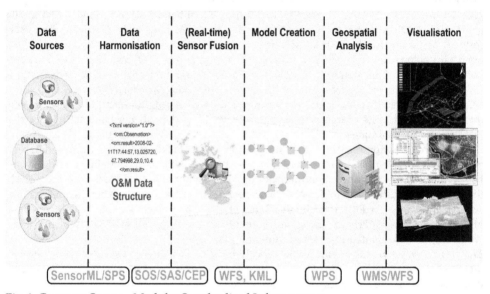

Fig. 1. Common Scents – Modular Standardised Infrastructure

According to principles of SOA and sustainable infrastructure development, we conceived data collection, processing and information provision architecture, which covers the whole process chain from sensor network development via measurement integration, data analysis and information visualisation, as shown in Figure 1. Thus, our approach builds the architectural bridge between domain-independent sensor network development and use case specific requirements for end user tailored information output for environmental monitoring purposes.

3.2 Standardised measurement infrastructure

The modules of the workflow shown in Figure 1 are separated by several interfaces, which are defined using open standards. The first central group of standards is subsumed under the term Sensor Web Enablement (SWE), an initiative by the OGC that aims to make sensors discoverable, query-able, and controllable over the Internet (Botts et al., 2007). Currently, the SWE family consists of seven standards comprising data models and service interfaces:

- *Sensor Model Language (SensorML)* – This standard provides an XML schema for defining the geometric, dynamic and observational characteristics of a sensor. Thus, SensorML assists in the discovery of different types of sensors, and supports the processing and analysis of the retrieved data, as well as the geo-location and tasking of sensors.
- *Observations & Measurements (O&M)* – O&M provides a description of sensor observations in the form of general models and XML encodings. This framework labels several terms for the measurements themselves as well as for the relationship between them. Measurement results are expressed as quantities, categories, temporal or geometrical values as well as arrays or composites of these.
- *Transducer Model Language (TML)* – Generally speaking, TML can be understood as O&M's pendant or streaming data by providing a method and message format describing how to interpret raw transducer data.
- *Sensor Observation Service (SOS)* – SOS provides a standardised web service interface allowing access to sensor observations and platform descriptions.
- *Sensor Planning Service (SPS)* – SPS offers an interface for planning an observation query. In effect, the service performs a feasibility check during the set up of a request for data from several sensors.
- *Sensor Alert Service (SAS)* – SAS can be seen as an event-processing engine whose purpose is to identify pre-defined events such as the particularities of sensor measurements, and then generate and send alerts in a standardised protocol format.
- *Web Notification Service (WNS)* – The Web Notification Service is responsible for delivering generated alerts to end-users by E-mail, over HTTP, or via SMS. Moreover, the standard provides an open interface for services, through which a client may exchange asynchronous messages with one or more other services.
- *Sensor Web Registry* – The registry serves to maintain metadata about sensors and their observations. In short, it contains information including sensor location, which phenomena they measure, and whether they are static or mobile. Currently, the OGC is pursuing a harmonisation approach to integrate the existing CS-W (Web Catalogue Service) into SWE by building profiles in ebRIM/ebXML (e-business Registry Information Model).

The functional connections between the SWE standards are illustrated in Figure 2.

It should be mentioned that the OGC is currently establishing the so-called *SWE Common* namespace specification, which aims at grouping elements that are used in more than one standard of the SWE family. In effect, this will minimise redundancy, and optimise re-usability and efficiency of the standards. SWE Common will mainly comprise very general elements such as counts, quantities, time elements or simple generic data representations.

More information on the Sensor Web Enablement initiative, the incorporated standards and the efforts to embed it into the OGC standard service development can be found on the OGC web site[1].

Aside from the SWE standard family, which is used throughout the sensor network process chain, other OGC standards are employed to build up the overall infrastructure. At first, the Web Processing Service (WPS) is used for integrating measurement data with thematic process models in order to generate contextual information layers. WPS (Schut, 2007) basically allows for the implementation and execution of pre-defined analysis processes

---

[1] http://www.opengeospatial.org

with dedicated input and output parameters. It supports synchronous and asynchronous data processing to enable sophisticated processing of large amounts of vector and raster data. The WPS standard has been discussed by Resch et al. (2010a) including issues such as input/output data definition, WPS profiling, asynchronous processing, and others.

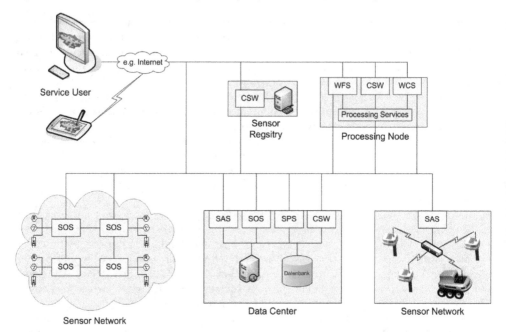

Fig. 2. Functional Connections Between the SWE Standards. (adapted from Botts et al., 2006)

The OGC developed a set of service interfaces for standardised data provision and visualisation dealing with various kinds of GIS data types. The Web Feature Service (WFS), the Web Map Service (WMS) and the Web Coverage Service (WCS) standards allow for access to geo-data such as vectors (point, line, polygon), raster images, and coverages (surface-like structures). More information about these standards and service implementations can be found on the OGC web site.

The essential benefit of using the OGC processing and data provision services mentioned above is the wide variety of standardised (GML, KML etc.) and custom output formats (GeoRSS, PDF etc.). This allows for the integration of the OGC service outputs into other processing, visualisation or decision support services including legacy COTS and open-source GIS analysis tools.

3.3 Data source: Geo-sensor web

In the Common Scents project, two pilot studies have been conducted. The first one used the *CitySense* sensing network (Murty et al., 2008) as the underlying sensing and data collection infrastructure. The main goal of the ongoing CitySense project is to build an urban sensor network to measure environmental parameters and is thus the data source for further data analysis. The project focuses on the development of a city-wide sensing system using an optimised network infrastructure. Currently, the network consists of 16 nodes deployed around the city of Cambridge measuring different environmental parameters such as CO_2 concentrations, air temperature, wind speed and direction, or precipitation. The final CitySense deployment will comprise up to 100 sensing nodes to build the basis for pervasive urban monitoring. (Resch et al., 2009b)

Another pilot experiment, which aimed at the deployment of a mobile sensor network, was conducted in the city of Copenhagen, Denmark. Ten bicycle mounted sensors[2] were used to collect environmental data (CO, NO_x, noise, air temperature and relative humidity) together with time and the geographic location using GPS – from which velocity and acceleration can be calculated. In this experiment of ubiquitous mobile sensing, we used the Sensaris City Senspod[3], a relatively low-cost sensor pod. The deployment in Copenhagen was a combined effort of the MIT SENSEable City Laboratory, and Københavns Kommune, Denmark.

To comply with the standardised infrastructure described in sub-section 3.1, we implemented several standardised services on top of these sensor networks, in accordance with the Live Geography approach (Resch et al., 2009a). For data access, we developed a Sensor Observation Service (SOS), which supplies measurement data in the standardised O&M format. It builds the O&M XML structure dynamically according to measured parameters and filter operations. To generate alerts e.g. in case of exceedance of a threshold, we implemented an XMPP (Extensible Messaging and Presence Protocol) based Sensor Alert Service (SAS). It is able to detect patterns and anomalies in the measurement data and generate alerts and trigger appropriate operations such as sending out an email or a text message, or to start a pre-defined GIS analysis operation.

3.4 Event-based alerting

Within the workflow described in sub-section 3.1, event recognition and processing happens in two different stages: 1.) at sensor level, Complex Event Processing (CEP) is used to detect errors in measurement values by applying different statistical operations such as standard deviations, spatial and temporal averaging, or outlier detection. Thus, it can be considered a mechanism for quality control and error prevention. To be able to detect condition changes in measurement values, we enhanced CEP and Event Stream Processing (ESP) techniques by the location parameter. Thus, these methods can also serve for the federal organisation of pre-defined geographical domain violations like geo-fences, and for tracing and analysing spatial patterns. 2.) after the data harmonisation process, CEP serves for spatio-temporal pattern recognition, anomaly detection, and alert generation in case of threshold exceedance.

Figure 3 shows the components of the CEP-based event processing component, which is built up in a modular structure.

Generally speaking, the event processing component serves as a connection between the data layer (sensor measurements) and the data analysis and data visualisation components, i.e. it prepares raw data in order to be process-able in the analysis and the visualisation layers. The first module is the data transportation, which connects different real-time and non-real-time data sources, i.e. it serves as an entry point into the event processing layer. Next, the retrieved data is passed on to the data transformation module, which prepares the data to be further processed. This 'processing' shall not be seen as data analysis, but more as data preparation. Basically, the transportation module converts the byte input stream to objects.

---

[2] http://senseable.mit.edu/copenhagenwheel
[3] http://www.sensaris.com

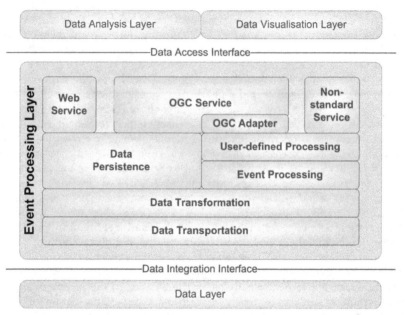

Fig. 3. CEP-based Event Processing Component Architecture

These objects can then be used by two higher-level components; firstly, by the data persistence component, which establishes a static data structure from the source data; secondly, by an event processing engine, which processes a real-time stream, identifies/selects events and pushes them to the user-defined processing module. The latter performs a kind of 'event filtering' and sends the resulting message to the service components.

One particularity, which shall be mentioned at this point, is the connection between the processing and the data persistence modules. The idea behind this functional link is that data, which have been prepared by the processing components, can either be pushed to the service components or they can be temporarily stored to be accessed by OGC services.

The two service-related components in Figure 3 (Web Service and Non-standard Service) serve as the direct interfaces to the data integration and data analysis layers. They offer the pre-processed raw data via a defined data structure, e.g. in a standardised output format such as GML, KML, geoTIFF etc. or in a custom output format.

For the OGC service component, all data are served over well-established open and standardised interfaces (OGC WFS, WMS and WCS). These XML web interfaces enable standardised data access and guarantee combinability of the various kinds of used data for further automated processing, as described in sub-section 3.2. In this way, pre-defined aggregation services can be implemented in the data analysis layer offering the results to a range of different users, i.e. platforms.

In addition to the standardised interfaces, also a non-standard service has to be created as existing OGC services don't support push mechanisms per se. A longer term option will be to replace these non-standard interfaces by push-capable standard services. More detailed information about the event processing component can be found in Resch et al. (2010b).

3.5 Real-time sensor fusion

We are currently facing a drastic increase in the availability of geospatial real-time data sources, and this applies especially to rapid developments and price reduction in sensing technologies. To make use of this immense amount of data within environmental monitoring systems, real-time data integration mechanisms have to be developed, which harmonise and fuse the different kinds of data. Furthermore, these data have to be provided in standardised formats in order to allow interoperability and collaboration between different institutions and data providers.

Most current data integration systems make use of a temporary database to combine different kinds of raw data, as stated in section 2. This approach has two distinctive disadvantages. Firstly, it manifests data into a physical structure and thus severely limits real-time capabilities. Secondly, the laborious operation of creating and filling a database table adds another step in the overall workflow, which decreases performance and expands implementation complexity and costs.

To overcome these shortcomings, we implemented the real-time data integration component in a custom data store for the open-source server GeoServer. This solution offers two main advantages: at first, data are fused on-the-fly in a highly dynamic, fast and parallelised process. At second, GeoServer provides standardised WFS, WMS and WCS outputs, as described above, which allows for simple integration into analysis and visualisation software. More about implementation details can be found in Resch et al. (2009a).

4. Results of spatio-temporal data analysis

Using the sensor web deployments described in section 3, we implemented two spatio-temporal data analysis modules.

The first pilot deployment has been carried out in the course of the *Copenhagen Wheel* project. This project was unveiled in Copenhagen on 15 December 2009 as part of 15th Conference of the Parties during the 2009 United Nations Climate Change Conference meeting. The Copenhagen Wheel is capturing information about carbon monoxide (CO), NO_x (NO + NO_2), noise, ambient temperature, relative humidity in addition to position, velocity and acceleration.

The environmental sensors were originally intended to be placed within the hub of the bicycle wheel however due to logistical pressure they were placed on bicycles ridden by couriers in Copenhagen going about their normal daily routine. Thus the testing was essentially a proof-of-concept. Ten cycles were instrumented and tested over 2 December 2009. It is believed that this was the first time multiple mobile sensors had been used in the field with such a large variety of environmental sensors.

The analysis component, which processes the collected data, performs a spatial Inverse Distance Weighting (IDW) interpolation (for a comparison with Kriging interpolation, s. Zimmermann et al., 1999) on temperature measurements, which will be used in further research efforts for correlation operations with emission distribution or traffic emergence, and for the detection of urban heat islands.

Moreover, the processing module analyses the CO distribution throughout the city of Copenhagen. The basic CO map containing the GPS traces and the output of the interpolation process – a navigable 3D map – are shown in Figure 4.

The second geo-processing component uses ArcGIS's Tracking Analyst tool to perform spatio-temporal analysis on measurement data over a period of time. In order to achieve a coarse overview of pollutant variability, we used CO_2 data captured by the CitySense network in Cambridge. This allows for correlating temporal measurement data fluctuation to traffic density, weather conditions or day-time related differences in a very flexible way. The lower left part of Figure 5 shows the temporal gradient of the measurement values. Running the time series then changes symbologies in the map on the right side accordingly in a dynamic manner in real-time.

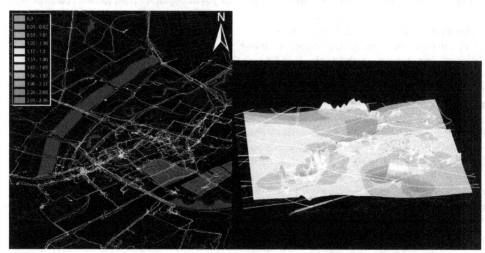

Fig. 4. Spatial Distribution of CO Values in the City of Copenhagen.

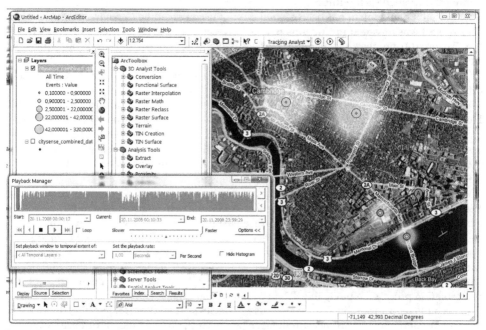

Fig. 5. Spatio-temporal CO_2 Data Analysis Using Tracking Analyst.

Figure 6 illustrates a time series representation of the measured parameters ambient temperature, CO, NO_x and noise. These measurements were taken in Copenhagen over a period of five hours on 2 December 2009. A first assessment shows that there are strong correlations between ambient temperature, CO and NO_x values.

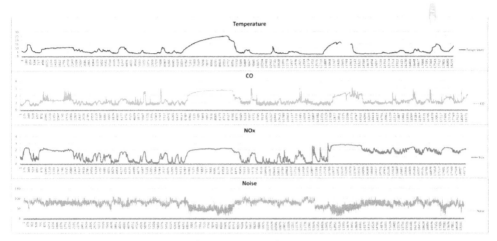

Fig. 6. Time Series Representation of Environmental Measurements.

Preliminary findings show that both CO and CO_2 are characterised by very high temporal and spatial fluctuations, which are induced by a variety of factors including temperature variability, time during the day, traffic emergence or 'plant respiration' – the fact that plants release major amounts of CO_2 over night. Also, CO is a measure of the efficiency of combustion in vehicles and may be used to reflect changing driving patterns and the sensitivity of air quality to larger scale environmental features such as the wind speeds over the city. However, the detailed interplay of these parameters still has to be investigated in a next step. Especially CO values measured in the Copenhagen pilot have to be normalised over humidity and temperature to perform further quantitative (absolute amounts) and qualitative (impact on public health) analysis.

5. Potential application areas

More than ten years ago, Al Gore articulated a vision of 'Digital Earth' as a multi-resolution, three-dimensional representation of the planet that would make it possible to find, visualise, and make sense of vast amounts of geo-referenced information on the physical and social environment (Gore 1998, for a comprehensive discussion see Craglia et al. 2008). Google Earth, NASA World Wind and other geo-browsers brought high resolution imagery to hundreds of millions of internet users and a major industry developed ways to explore data geographically, and visualise overlaid information provided by both the public and private sectors, as well as citizens who volunteer new data (Goodchild, 2007).

Generally speaking, fine-grained urban sensing greatly enhances our knowledge of the environment by adding objective and non-visible data layers in real-time (Resch et al., 2008). In other words, these systems help us increase our capacity to observe and understand the city, and the impacts on and by society. This seems to be a very desirable state because more accurate data about local air temperature, atmospheric humidity, gaseous and particulate air pollution, and traffic emissions can positively influence areas such as public health, traffic management or emergency response. Apart from this information enrichment, accurate sensor measurements also have a much broader influence: considering for example that 'air quality' is only a surrogate for the effects of pollutants on humans makes a fine-grained air quality map a very sensitive information layer, as discussed in section 4.

Within the Common Scents project, we focus on the use case of air quality monitoring for use in the public health sector. However, we designed the monitoring infrastructure in such a modular way that it is not bound to one single application area. Below, several practically motivated fields of real-world applications are described, which could use the same infrastructure presented in section 3.

Public Health has been asked to participate in policymaking on 'quality of life' issues increasingly over the past decade. The superimposing of the medical model to describe the impact of conditions that have traditionally been regarded as nuisances has created a great challenge, particularly in the field of environmental health.

One pollutant often used to serve as a proxy is NO_x, which technically represents various gaseous species comprised of oxygen and nitrogen molecules. Another indicator of near-roadway effects that has gained recent attention is ultrafine particulates (UFPs), particles that are less than 0.1 microns (100 nm) in diameter. Thus, air quality measurements of hazardous air pollutants can be widely associated with traffic (non-point sources). A pervasive sensor network could help capture measurements in high spatial and temporal resolution to take short-term measures (dynamically adapt traffic management or send out alerts to citizens in case of threshold exceedance). Also, it could support traditional long-term studies on the impact of certain pollutants on public health.

The use case of noise mapping has received a lot of attention recently. Many disputes within the research field emerge from noise impacts associated with construction, excavation or some other commercial or industrial enterprise. These disputes also arise from use of domestic landscaping equipment, like leaf blowers and snow blowers. The limits imposed by the city on noise generation are intended to assess the background noise levels. A source cannot be held responsible for noise levels that exceed the city's allowable limits if the ambient noise in that area already exceeds those limits. The development of noise 'maps' may not immediately result in satisfaction from aggrieved residents, but it can be used to consider the noise impact of future development and zoning policies. It may also contribute to efforts to reduce the number of cars travelling across the city by adding the noise impact dimension to the discussion. This is much more likely to be given full consideration if it can be demonstrated with highly resolved data maps, which can be generated in near real-time using the Common Scents infrastructure.

The urban heat island effect describes the contribution of the built environment to the ambient temperature within urban areas. While this is not likely to become a primary public health concern, it has great bearing on efforts to limit the loss of heating energy across the city. Different agencies have been established to work on a long-term strategy to reduce overall energy use (e.g. Cambridge Energy Alliance: http://www.cambridgeenergyalliance.org) and to encourage individual homeowners and building owners to evaluate their energy loss. It is quite possible that small changes in heat loss, as described through a detailed heat map of the city over time, could show progress towards energy efficiency in a materials way. This could be used both as an evaluation tool in tracking the city's progress, and as a means to engage the public in the energy goals of the community.

6. Conclusion

Ubiquitous and continuous environmental monitoring is an enormous challenge, and this is particularly true in the urban context, which poses very specific challenges as well technologically as socially and politically. In this chapter we discussed several of these issues, and outlined how our approach can meet future requirements for urban sensing.

The focus is to contribute to a 'complete' picture of a living city for decisions makers, planners and operators beyond locational analysis. This may be seen distinct from a number of citizen-centred sensor approaches and context-aware systems. While a number of people-centric pervasive sensing systems are notable successes (Campbell et al. 2006), most of these examples focus on localising people and objects in a defined environment to enable context aware applications. In such projects, the notion of sensing is confined to supporting location-based context-awareness. In our *Live Geography* approach a more general integrated sensing architecture to support the diversity of applications and hardware platforms has been developed.

Based on the Live Geography approach, we outlined the Common Scents concept, which tries to establish an interoperable, modular and flexible sensing and data analysis infrastructure, as opposed to hitherto monolithic sensor networks. To prove our system's portability, we did implementations in two different pilot deployments (Cambridge, MA US and Copenhagen, Denmark) using the same data integration and analysis infrastructure.

Further exploitation of this approach is planned for other cities. We see more future challenges in the socio-political domain rather in the technological development necessary. It becomes more and more obvious that a cross-disciplinary group of researchers and technologists needs to persistently interact with end users. Only then we may achieve a wide appreciation of sensing which is needed to support future civic, cultural, and community life in cities. In many parts of the world, notably Germany and some Western European countries, attempts to 'completely' map cities are very sensitive. Google faces great problems with its StreetView approach. An integrated Common Scents must provide a clearly recognisable benefit to the citizens in order to be appreciated by all societal groups. Public Health applications may have a good chance to get accepted although some of the capabilities, for instance the ability to measure remotely the conditions of people in real time, raise social concerns centred on privacy issues. Methods for sensor data fusion and designs for human-computer interfaces are both crucial for the full realisation of the potential of integrated and pervasive sensing.

We also believe that the impact of pervasive sensing in the city has to be carefully assessed. We found that e.g. providing very fine-grained information layers might on the one hand be a powerful decision support instrument, but on the other hand too detailed environmental information might also have negative effects. As the term 'air quality' is just a surrogate for more personal impacts such as life expectation or respiration diseases, this information could yield a very broad impact in various kinds of areas such as housing market, the insurance sector or urban planning in general.

As the Common Scents concept has been developed and implemented together with the Public Health Department of the City of Cambridge, MA US as concrete end users, we believe that our approach can respond to dedicated needs of the city management. Therefore, the longer-term goal is to enhance people's perception of their environment by adding unseen information layers and thus changing their short-term behaviour by providing real-time decision support.

7. Acknowledgement

The Common Scents project is a concerted effort between the Research Studio iSPACE, the MIT SENSEable City Lab, the City of Cambridge's Public Health Department and the Harvard University School of Engineering and Applied Sciences. We would like to thank all internal and external collaborators for making this project happen.

Several technical parts of this research have been funded through the European Commission in the course of the FP7 GENESIS project. Furthermore, the Austrian Ministry of Science and Research (BMWF) funded pieces of the efforts in the Research Studio iSPACE.

The research described in this project by one of the authors (REB) was funded in part by the Singapore National Research Foundation (NRF) through the Singapore-MIT Alliance for Research and Technology (SMART) Center for Environmental Sensing and Monitoring (CENSAM).

The Copenhagen Wheel team from the SENSEable City Laboratory, MIT is composed of Christine Outram (Project Leader), Rex Britter, Andrea Cassi, Xiaoji Chen, Jennifer Dunnam, Paula Echeverri, Myshkin Ingawale, Ari Kardasis, E Roon Kang, Sey Min, Assaf Biderman and Carlo Ratti The courier cyclists were organized by Signe Gaarde in Copenhagen.

8. References

Balazinska, M., Deshpande, A., Franklin, M.J., Gibbons, P.B., Gray, J., Hansen, M., Liebhold, M., Nath, S., Szalay, A., Tao, V. (2007) Data Management in the Worldwide Sensor Web. IEEE Pervasive Computing, vol. 6, no. 2, pp. 30-40, Apr-Jun, 2007.

Botts, M., Robin, A., Davidson, J. and Simonis, I. (Eds.) (2006) OpenGIS Sensor Web Enablement Architecture Document. http://www.opengeospatial.org, OpenGIS Interoperability Project Report OGC 06-021r1, Version 1.0, 4 March 2006. (12 June 2011)

Botts, M., Percivall, G., Reed, C., and Davidson, J. (Eds.) (2007) OGC Sensor Web Enablement: Overview and High Level Architecture. http://www.opengeospatial.org, OpenGIS White Paper OGC 07-165, Version 3, 28 December 2007. (17 June 2011)

Campell, A., Eisenman, S.B., Lane, N.D., Miluzzo and E., Peterson, R.A. (2006) People-Centric Urban Sensing. Proceedings of the 2nd annual international Workshop on Wireless Internet, Boston, 2006.

Center for Coastal and Land-Margin Research (2009) CORIE. http://www.ccalmr.ogi.edu/CORIE. (14 June 2011)

Goodchild, M. (2007). Citizens as sensors: The world of volunteered geography. Geojournal, 69: 211-221.

Gore, A. (1998) The Digital Earth: Understanding Our Planet in the 21st Century. Speech given by Vice President Al Gore at the California Science Center, Los Angeles, California, on January 31, 1998, http://www.isde5.org/al_gore_speech.htm. (27 June 2011)

Gross, N. (1999) 14: The Earth Will Don an Electronic Skin. http://www.businessweek.com, BusinessWeek Online, 30 August 1999. (20 May 2011)

Harrie, L. (2004) Using Simultaneous Graphic Generalisation in a System for Real-Time Maps. Papers of the ICA Workshop on Generalisation and Multiple Representation, August 20-21, 2004, Leicester (electronic version available at http://ica.ign.fr/Leicester/paper/Harrie-v2-ICAWorkshop.pdf).

Kansal, A., Nath, S., Liu, J. and Zhao, F. (2007) SenseWeb: An Infrastructure for Shared Sensing. IEEE Multimedia. vol. 14, no. 4, pp. 8-13, October-December 2007.

Lehto, L. and Sarjakoski, L.T. (2005) Real-time Generalisation of XML-encoded Spatial Data for the Web and Mobile Devices. International Journal of Geographical Information Science, vol. 19, no. 8-9, pp. 957-973.

Murty, R., Mainland, G., Rose, I., Chowdhury, A., Gosain, A., Bers, J. and Welsh, M. (2008) CitySense: A Vision for an Urban-Scale Wireless Networking Testbed. In Proceedings of the 2008 IEEE International Conference on Technologies for Homeland Security, Waltham, MA, May 2008.

Nagel, D. (2003) Pervasive Sensing. Proceedings of the SPIE, vol. 4126, no. 71, doi:10.1117/12.407543, 2003.

Nath, S., Liu, J. and Zhao, F. (2006) Challenges in Building a Portal for Sensors World-Wide. First Workshop on World-Sensor-Web: Mobile Device Centric Sensory Networks and Applications (WSW'2006), Boulder CO, 31 October, 2006.

National Oceanic and Atmospheric Administration (2011) nowCOAST: GIS Mapping Portal to Real-Time Environmental Observations and NOAA Forecasts. http://nowcoast.noaa.gov. (15 June 2011)

Paulsen, H. and Riegger, U. (2006). SensorGIS - Geodaten in Echtzeit. In: GIS-Business, vol. 8/2006, pp. 17-19, Cologne.

Paulsen, H. (2008) PermaSensorGIS – Real-time Permafrost Data. In: Geoconnexion International Magazine, vol. 02/2008.

Rahm, E., Thor, A. and Aumueller, D. (2007) Dynamic Fusion of Web Data. XSym 2007, Vienna, Austria, pp.14-16.

Resch, B., Calabrese, F., Ratti, C. and Biderman, A. (2008) An Approach Towards a Real-time Data Exchange Platform System Architecture. Sixth Annual IEEE-IARIA International Conference on Pervasive Computing and Communications, Hong Kong, 17-21 March 2008.

Resch, B., Mittlboeck, M., Girardin, F., Britter, R. and Ratti, C. (2009a) Live Geography – Embedded Sensing for Standardised Urban Environmental Monitoring. International Journal on Advances in Systems and Measurements, 2(2&3), ISSN 1942-261x, pp. 156-167.

Resch, B., Mittlboeck, M., Lipson, S., Welsh, M., Bers, J., Britter, R. and Ratti, C. (2009b) Urban Sensing Revisited – Common Scents: Towards Standardised Geo-sensor Networks for Public Health Monitoring in the City. In: Proceedings of the 11th International Conference on Computers in Urban Planning and Urban Management - CUPUM2009, Hong Kong, 16-18 June 2009.

Resch, B., Sagl, G., Blaschke, T. and Mittlboeck, M. (2010a) Distributed Web-processing for Ubiquitous Information Services - OGC WPS Critically Revisited. In: Proceedings of the 6th International Conference on Geographic Information Science (GIScience2010), Zurich, Switzerland, 14-17 September 2010.

Resch, B., Lippautz, M. and Mittlboeck, M. (2010b) Pervasive Monitoring - A Standardised Sensor Web Approach for Intelligent Sensing Infrastructures. Sensors - Special Issue 'Intelligent Sensors 2010', 10(12), 2010, pp. 11440-11467.

Rittman, M. (2008) An Introduction to Real-Time Data Integration. http://www.oracle.com/technology/pub/articles/rittman-odi.html, 2008. (22 May 2011)

Riva, O. and Borcea, C. (2007) The Urbanet Revolution: Sensor Power to the People!. IEEE Pervasive Computing, vol. 6, no. 2, pp. 41-49, Apr-Jun, 2007.

Sarjakoski, T. et al. (2004) Geospatial Info-mobility Service by Real-time Data-integration and Generalisation. http://gimodig.fgi.fi. (22 June 2011)

Schut, P. (ed.) (2007) Web Processing Service. OpenGIS Standard, version 1.0.0, OGC 05-007r7, 8 June 2007.

Srivastava, M., Hansen, M., Burke, J., Parker, A., Reddy, S., Saurabh, G., Allman, M., Paxson, V. and Estrin D. (2006). Wireless Urban Sensing Systems. Technical Report 65, Center for Embedded Network Sensing, UCLA, April 2006.

Sybase Inc. (2008) Real-Time Events Data Integration Software.
http://www.sybase.com/products/dataintegration/realtimeevents. (07 July 2011)

University of Oklahoma (2009) OKCnet. http://okc.mesonet.org. (12 March 2011)

Xu, N. (2004) A Survey of Sensor Network Applications. http://courses.cs.tamu.edu, Computer Science Department, University of Southern California, 2004. (10 July 2011)

Zimmerman, D., Pavlik, C., Ruggles, C. and Armstrong, M.P. (1999) An Experimental Comparison of Ordinary and Universal Kriging and Inverse Distance Weighting. Mathematical Geology, 31(4), pp. 375-390.

8

Collaborative Environmental Monitoring with Hierarchical Wireless Sensor Networks

Qing Ling[1], Gang Wu[1] and Zhi Tian[2]
[1]*Department of Automation, University of Science and Technology of China*
[2]*Department of Electrical and Computer Engineering, Michigan Technological University*
[1]*China*
[2]*USA*

1. Introduction

In the last decade, advances in wireless communication and micro-fabrication have motivated the development of large-scale wireless sensor networks (Akyildiz et al., 2002; Yick et al., 2008). A large number of low-cost sensor nodes, equipped with sensing, computing, and communication units, organize themselves into a multi-hop network. The wireless sensor network takes measurements from the environment, processes the sensory data, and transmits the sensory data to end-users. Beginning from the seminar work in (Estrin et al., 1999; 2002), the wireless sensor network technology has been well recognized as a revolutionary one that transforms everyday life. Typical applications of wireless sensor networks include military target tracking and surveillance (Simon et al., 2004; He et al., 2006), precise agriculture (Langendoen et al., 2006; Wark et al., 2007), industrial automation (Gungor and Hancke, 2009), structural health monitoring (Li and Liu, 2007; Ling et al., 2009), environmental and habitat monitoring (Zhang et al., 2004; Corke et al., 2010), to name a few.

1.1 Network infrastructure

To organize the large amount of sensor nodes and enable efficient data collection, a wireless sensor network generally adopts one of the following three infrastructures: centralized, decentralized, and hierarchical. In the centralized infrastructure, sensor nodes transmit the sensory data to the fusion center via multi-hop communication. In the decentralized infrastructure, each sensor node firstly refines the sensory data through collaborative and decentralized in-network processing with the neighboring sensor nodes, and secondly transmits the refined data to the fusion center. While in the hierarchical infrastructure, sensor nodes are divided into multiple clusters, and sensor nodes within one cluster send their sensory data to the cluster head. These cluster heads either transmit the collected sensory data to the fusion center, or collaboratively process them and transmit the refined one to the fusion center. These two different implementations of the hierarchical infrastructure, centralized processing and decentralized collaboration, are depicted in Figure 1.

In deploying a wireless sensor network, the choice of its infrastructure is decided by several key factors: energy, bandwidth, robustness, etc. Sensor nodes are often equipped with batteries and recharging is difficult. Since wireless data transmission is the main source of energy consumption of a sensor node (Sadler, 2005), the network infrastructure

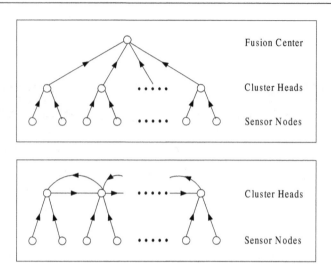

Fig. 1. Two different schemes of implementing the hierarchical infrastructure: (TOP) centralized processing in a fusion center and (BOTTOM) decentralized collaboration among cluster heads.

should guarantee that each sensor node has low data transmission rate while successfully accomplishing the data collection task. Bandwidth is also a kind of precious resource in wireless environment; over-competition of wireless channels leads to frequent retransmission and hence consumes more energy. Further, sensor nodes are often fragile due to being out of batteries or other physical damages. The network infrastructure should be carefully designed such that the failure of few sensor nodes shall not result in the malfunction of the whole network.

When the network size is small, the centralized infrastructure is an acceptable choice. Take a volcano monitoring network containing 3 sensor nodes (Werner-Allen et al., 2005) as an example, these sensor nodes directly connect to a fusion center which collects sensory data and transmits them to the end-user. Later on the network is extended to the scale of 16 sensor nodes (Werner-Allen et al., 2006), and the sensor nodes communicate with the fusion center via multi-hop relays. However, for GreenOrbs (Liu et al., 2011), a large-scale forest monitoring network composed of up to 330 sensor nodes, experiments demonstrate that sensor nodes within some "hot areas" may face higher competition for bandwidth, consume more energy, and be more sensitive to the failure of sensor nodes. The decentralized infrastructure, on the other hand, has great potential to reduce the total amount of transmitted data and hence improve the energy efficiency via in-network collaboration; further, it also enhances robustness of the network since all sensor nodes play equal roles (Ling and Tian, 2010). Nevertheless, collaboration of the sensor nodes brings more difficulty to network coordination, and is subject to the limited processing and communication capabilities of sensor nodes. For this reason, the decentralized infrastructure is still far from practical applications. To the best of our knowledge, most large-scale wireless sensor networks are deployed with the hierarchical infrastructure. Following we give some examples: ExScal, an intrusion detection network with more than 1000 sensor nodes and more than 200 backbone nodes (Arora et al., 2005); VigilNet, a military surveillance network with 200 sensor nodes (He et al., 2006); Trio, a target tracking network with 557 solar-powered sensor nodes

(Dutta et al., 2006); SenseScope, an environmental monitoring network consisting of from 3 to 97 sensor nodes (Barenetxea et al., 2008). In view of this fact, we will focus on the design of a hierarchical wireless sensor network.

1.2 Our contributions

In some hierarchical wireless sensor networks such as ExScal (Arora et al., 2005), the cluster heads are specifically designed, having better data processing and wireless communication abilities than general sensor nodes, and equipped with stronger or even uninterruptible power sources. These cluster heads can directly transmit the collected data to a remote fusion center, without introducing any collaborative processing among cluster heads. However, in most wireless sensor networks, cluster heads are elected from sensor nodes to simplify design, deployment, and maintenance. For example, in the LEACH protocol (Heinzelman et al., 2002), sensor nodes autonomously elect cluster heads, aiming at evenly distributing energy consumption among all sensor nodes so that there are no overly-utilized sensor nodes that will run out of energy before the others. In this case, how to process the collected sensory data in the cluster heads is a critical problem to accomplishing the data collection task while maximizing the network lifetime.

This chapter addresses this problem; specifically, we study a generalized environmental monitoring model with large-scale hierarchical wireless sensor networks, and focus on two questions: for cluster heads in a hierarchical network, *should they collaborate or not collaborate* and *how can they collaborate*. Our contributions are two-fold.

First, through theoretical analysis and simulation validation, we make the following recommendations on whether to collaborate or not: when each cluster head has a large amount of data to process (namely, each cluster contains a large number of sensor nodes) and multi-hop relay is necessary to communicate with a fusion center (namely, cluster heads have limited communication range), decentralized data processing among cluster heads is more efficient; otherwise centralized decision-making with the aid of a fusion center can be advantageous.

Previous work, such as (Rabbat and Nowak, 2004; Aldosari and Moura, 2004), has suggested similar network design principles in the context of decentralized infrastructures: when each sensor node collects a large amount of data or the size of the network is large, collaborative processing is more efficient than centralized decision-making. This paper extends the conclusions to hierarchical networks, and compares decentralized versus centralized processing among cluster heads rather than among all sensor nodes.

Second, we develop a decentralized collaborative algorithm for decision making among the sub-network of cluster heads, after they have collected sensory data from local sensor nodes within their individual clusters. Particularly, we study a typical environment monitoring application, in which a large-scale hierarchical wireless sensor network is deployed to monitor sparsely occurring phenomena over a large sensing field. The monitoring problem is formulated as a non-negative quadratic program, which optimizes a sparse decision vector depicting the spatial map of the phenomena of interest. An optimal iterative algorithm, in which cluster heads iteratively exchange information and make decisions, is proposed based on the alternating direction method of multipliers (ADMM) (Bertsekas and Tsitsiklis, 1997). Our development is permeated with the benefits of compressive sensing (Donoho et al., 2006). Exploiting the sparse nature of the unknown phenomena, we allow the number of sensor nodes to be much smaller than what would have been required in a traditional scheme for sensing at high spatial resolution over a large field. In this sense, our proposed algorithm is also applicable to other compressive sensing problems in distributed systems.

1.3 Chapter organization

The rest of this chapter is organized as follows. We first give a brief survey on the applications of wireless sensor networks in environmental monitoring. Second, we study a generalized environmental monitoring model with large-scale hierarchical wireless sensor networks and develop a decentralized collaborative algorithm for decision making among the cluster heads. Finally we discuss the design consideration, namely, to collaborate or not to collaborate, based on theoretical analysis and simulation results.

2. A brief survey

In this section, we give a brief survey on the applications of wireless sensor networks in environmental and habitat monitoring. Though this overview is far from complete, it reflects the promising future of the wireless sensor network technology in helping us understand and protect natural environment.

For environmental and habitat monitoring applications, one of the first known practical wireless sensor networks was deployed by a group at Berkeley in 2002, on Great Duck Island on the coast of Maine, USA. Two networks with a total of 147 sensor nodes collect data to study the ecology of the Leach's Storm Petrel (Szewsczyk et al., 2004). Later on, the Macroscope system which contains 33 sensor nodes, also developed at Berkeley, was used for microclimate monitoring of a redwood tree (Tolle et al., 2005). Another notable application is ZebraNet, which used GPS technology to record position data in order to track long term animal migrations. In the prototype system, researchers deployed 7 sensor nodes on zebras in Kenya (Zhang et al., 2004). Energy harvesting technologies have also attracted much research interest to address the challenge of energy supply in remote environmental monitoring applications. One successful example is LUSTER, which was developed at University of Virginia, featuring a specifically designed hybrid multichannel energy harvesting device (Selavo et al., 2007). Accompanied with the unprecedented data collection opportunities, data processing also emerges as a new challenge in the wireless sensor network technology. The data processing task is indeed application-oriented. For example, an ellipsoids-based anomaly detection algorithm was designed to monitor unusual and anomalous behaviors in a particular marine ecosystem (Bedzek et al., 2011). The network was deployed in 2009 at the Heron Island, Australia, as part of the Great Barrier Reef Ocean Observation System.

One significant advantage of wireless sensor networks over traditional data collection techniques is that they can be applied in harsh environments. For example, in the GlacsWeb system, researchers at University of Southampton deployed 9 sensor nodes inside a glacier (Martinez et al., 2004). The sensor nodes monitored pressure, temperature, and tilt, in order to monitor subglacial bed deformation. Even on active volcanos, which are often forbidden areas for data collection, wireless sensor networks can still work well. In the work of (Werner-Allen et al., 2005; 2006), one small sensor network with 3 sensor nodes was deployed on Vlcan Tungurahua in Ecuador as a proof of concept in 2004; then in 2005, the network size was extended to 16 sensor nodes. Wireless sensor networks are also fit for aquatic environmental monitoring applications. In (Alippi et al., 2011), a robust, adaptive, and solar-powered network was developed in 2007 for such an application. The network was deployed in Queensland, Australia, for monitoring the underwater luminosity and temperature, information necessary to derive the health status of the coralline barrier. At the same time, sensory data can be used to provide quantitative indications related to cyclone formations in tropical areas.

However, applying wireless sensor networks in environmental monitoring is still a challenging task when the network size is large. When the number of sensor nodes increases, difficulties emerge for system integration (creating an end-to-end system that delivers data to the end-user), performance (reliability, accuracy, and calibration), productivity (how well the sensory data assists the end-user and how to reduce the total cost in implementing the wireless sensor network), etc (Corke et al., 2010). One negative example is reported in (Langendoen et al., 2006), in which researchers at Delft University of Technology deployed a large-scale network in a potato field to improve the protection of potatoes against disease. The application was not successful due to unanticipated issues; nevertheless, the lessons are precious, such as software, hardware, and even team coordination. A systematic discussion, named as "the hitchhiker's guide", is provided in (Barenetxea et al., 2008). Based on the deployment of a wireless sensor network on a rock glacier located at a mountain in the Swiss Alps, this guide covers almost all stages of a project, from hardware and software development, testing and preparation, to deployment. One of the recent efforts to investigate the practical implementations of large-scale wireless sensor networks is the GreenOrbs system (Liu et al., 2011). The network with 330 sensor nodes was deployed in Tianmu Mountian, China, aiming at all-year-around ecological surveillance in the forest. It is shown that many traditional design guidelines for small-scale wireless sensor networks can be questionable for large-scale applications.

3. Problem formulation

In this chapter, we focus on a generalized event detection model for environmental monitoring applications. Let us consider a large-scale wireless sensor network randomly deployed in a two-dimensional area for monitoring sparsely occurring events. The network has a set of L sensor nodes, denoted as $\mathcal{L} = \{v_l\}_{l=1}^{L}$. Sensor nodes are divided into I clusters, each having one cluster head in the set $\mathcal{I} = \{c_i\}_{i=1}^{I}$. Sensor nodes within a cluster are able to directly transmit measurements to the cluster head, and the cluster head is aware of the positions of all sensor nodes within its cluster. Further, the cluster heads have a common communication range r_C such that the sub-network of cluster heads is bi-directionally connected.

3.1 Basic assumptions

Suppose that at each sampling time, multiple phenomena may occur in the sensing field. Our objective is to detect and identify the source locations and estimate their amplitudes from sensory measurements. We make the following basic assumptions for the sensing problem of interest, similar to those in (Bazerque and Giannakis, 2010):

(A1): The sensing field is viewed through a spatial grid with K grid points denoted by $\mathcal{K} = \{g_k\}_{k=1}^{K}$, whose locations are known to the corresponding cluster heads. Each event can occur only at a grid point, indicating the spatial resolution offered by this sensor network. The amplitude of an event occurring at grid point g_k is x_k.

(A2): The influence of a unit-amplitude event at grid point g_k on a sensor point v_l is f_{kl}. Generally speaking, f_{kl} is decided by the distance d_{kl} between g_k and v_l.

(A3): The measurement of one sensor node is the linear superposition of the influences of all phenomena plus random noise. Mathematically, the measurement b_l of sensor node v_l is hence $b_l = \sum_{k=1}^{K} f_{kl} x_k + e_l$ in which x_k is the amplitude of event at $g_k \in \mathcal{K}$ and e_l is measurement noise.

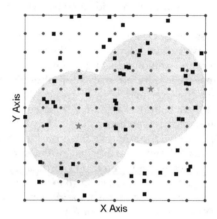

Fig. 2. The sensor nodes denoted as solid squares are uniformly randomly deployed in the monitoring area. The candidate positions for phenomena are grid points denoted as solid dots. Phenomena denoted as pentagrams occur at the current snapshot, and the shadow regions illustrate the influence of phenomena.

These assumptions are depicted in Figure 2. The sensor nodes denoted as solid squares are uniformly randomly deployed in the monitoring area. The candidate positions for phenomena are grid points denoted as solid dots. Phenomena denoted as pentagrams occur at the current snapshot, and the shadow regions illustrate the influence of phenomena.

The assumption **(A1)** simplifies the recovery problem by confining the sources of phenomena to grid points. Without this assumption, an alternative way is to use positions and amplitudes of the sources as decision variables and formulate a least squares problem. However, this formulation is highly nonlinear and intractable, since the number of decision variables is even unknown. Based on **(A1)**, we can formulate the otherwise nonlinear problem as recovering the vector $\mathbf{x} = [x_1, ..., x_K]^T$ from linear measurements $b_l, \forall l$. Entries in $\mathbf{x}$ with nonzero values reveal the locations and amplitudes of the multiple phenomena of interest. This assumption approximately holds when the grid points are dense; namely, the density of the grid points decides the spatial resolution of the recovery algorithm.

The assumption **(A2)** describes the influence of one event on the entire sensing field. For example, in target tracking or nuclear radioactive detection, the influence of a source decreases polynomially as the distance increases. Without loss of generality, we define the influence function as $f_{kl} = \exp(-d_{kl}^2/\sigma^2)$ for grid point g_k and sensor point v_l, where σ is a common constant. This Gaussian-shaped function well approximates the influence of many practical events.

Based on the assumption **(A3)**, we readily have the following least squares formulation for recovering $\mathbf{x}$:

$$\min \quad \sum_{v_l \in \mathcal{L}} (b_l - \sum_{k=1}^{K} f_{kl} x_k)^2, \tag{1}$$

or equivalently in a matrix form:

$$\min \quad ||\mathbf{F}\mathbf{x} - \mathbf{b}||_2^2. \tag{2}$$

Here $\mathbf{b} = [b_1, ..., b_L]^T$ is the measurement vector and $\mathbf{F}$ is the $L \times K$ influence matrix with its l-th row given by $[f_{1l}, ..., f_{Kl}]$.

Nevertheless, the least squares formulation (2) ignores the sparsity of the vector $\mathbf{x}$. Notice that when the grid is dense, the number of events is generally much smaller than the number of grid points; hence the vector $\mathbf{x}$ is a sparse vector with a large amount of zero entries. Without considering this prior knowledge, the least squares formulation (2) leads to a non-sparse solution, which means a non-neglectable number of false alarms. The sparsity of a signal vector can be measured by its ℓ_1 norm (Donoho et al., 2006). Exploiting the sparse nature of $\mathbf{x}$ to alleviate false alarms, we formulate the following ℓ_1 regularized least squares problem (Kim et al., 2007):

$$\min \quad \frac{\lambda}{2}||\mathbf{Fx} - \mathbf{b}||_2^2 + ||\mathbf{x}||_1. \tag{3}$$

Here λ is a non-negative weight.

3.2 Decentralized optimization

In a centralized setting, the ℓ_1 regularized least squares problem (3) has been extensively studied in both signal processing and numerical optimization communities (Donoho et al., 2006; Figueiredo et al., 2007). However, in a large-scale wireless sensor network, centralized processing is not efficient in terms of energy consumption and communication overhead. In contrast, collaborative signal processing among cluster heads is preferred, leading to a robust and scalable network.

We address this issue by developing a collaborative sparse signal recovery algorithm in the chapter. Sensor nodes or cluster heads do not necessarily exchange information with a fusion center; rather, sensor nodes only need to transmit measurements to their cluster heads, and cluster heads iteratively optimize the decision vector $\mathbf{x}$ via exchanging information with their neighboring cluster heads.

For each cluster head c_i, let us collect the local measurements $\{v_l\}$ within this cluster and their corresponding l-th rows in the measurement matrix $\mathbf{F}$ into a sub-vector $\mathbf{b}_i$ and a sub-matrix $\mathbf{F}_i$, for all sensor nodes v_l whose cluster head is c_i. Per assumptions (A1) and (A2), each cluster head knows all sensor node locations and grid point locations within its cluster, which means that $\mathbf{F}_i$ is known to c_i. Hence the problem boils down to the following one: *suppose that the local measurement vector $\mathbf{b}_i$ and corresponding measurement matrix $\mathbf{F}_i$ are available to each cluster head c_i, $\forall i$, how can we design a decentralized algorithm to recover the signal $\mathbf{x}$ via collaboration among the cluster heads?*

Let $\mathbf{x}_i$ denote the local copy of the decision vector $\mathbf{x}$ at c_i, $\forall c_i \in \mathcal{I}$. Meanwhile, given the communication range r_C, the set of neighboring cluster heads of c_i is denoted by $\mathcal{N}_i$, with cardinality $|\mathcal{N}_i|$. The formulation in (3) can be transformed to the following consensus optimization problem:

$$\min \quad \sum_{i=1}^{I}(\frac{\lambda}{2}||\mathbf{F}_i\mathbf{x}_i - \mathbf{b}_i||_2^2 + \frac{1}{I}\mathbf{1}^T\mathbf{x}_i),$$
$$s.t. \quad \mathbf{x}_i = \mathbf{x}_j, \quad \forall c_i \in \mathcal{I}, c_j \in \mathcal{N}_i. \tag{4}$$

The $K \times 1$ all-one vector $[1, 1, ..., 1]^T$ is denoted as $\mathbf{1}$. Here, cluster heads optimize their own local copies of $\mathbf{x}$ separately, and these decision vectors are forced to be equal via the consensus constraints. An alternative formulation is to force $\mathbf{x}_i$ to consent with the average of its neighboring decisions, as follows:

$$\min \quad \sum_{i=1}^{I}(\frac{\lambda}{2}||\mathbf{F}_i\mathbf{x}_i - \mathbf{b}_i||_2^2 + \frac{1}{I}\mathbf{1}^T\mathbf{x}_i),$$
$$s.t. \quad \mathbf{x}_i = \frac{1}{|\mathcal{N}_i|}\sum_{c_j \in \mathcal{N}_i} \mathbf{x}_j, \quad \forall c_i \in \mathcal{I}. \tag{5}$$

It has been proved that if the sub-network of the cluster heads is bi-directionally connected, then (4) and (5) are equivalent to (3) (Zhu et al., 2007). Both (4) and (5) can be solved similarly, as below.

4. Collaborative environmental monitoring algorithm

We now apply an optimal algorithm, the alternating direction method of multipliers (ADMM) (Bertsekas and Tsitsiklis, 1997), to solve (4).

4.1 Algorithm development

To solve (4) with the ADMM, we first introduce a new block of auxiliary variables. Then (4) can be rewritten as:

$$\min_{\{x_i\},\{z_{ij}\}} \quad \sum_{i=1}^{I}(\frac{\lambda}{2}||F_i x_i - b_i||_2^2 + \frac{1}{I}1^T x_i),$$
$$\text{s.t.} \quad x_i = z_{ij}, x_j = z_{ij}, \quad \forall c_i \in \mathcal{I}, c_j \in \mathcal{N}_i. \tag{6}$$

Here z_{ij} is an auxiliary vector attached to x_i and x_j. The augmented Lagrangian function of (6) is:

$$L_a\left(\{x_i\}, \{z_{ij}\}, \{\beta_{ij}\}, \{\gamma_{ij}\}\right) = \sum_{i=1}^{I}(\frac{\lambda}{2}||F_i x_i - b_i||_2^2 + \frac{1}{I}1^T x_i)$$
$$+ \sum_{i=1}^{I}\sum_{c_j \in \mathcal{N}_i} \beta_{ij}^T(x_i - z_{ij}) + \frac{d}{2}\sum_{i=1}^{I}\sum_{c_j \in \mathcal{N}_i}||x_i - z_{ij}||_2^2 \tag{7}$$
$$+ \sum_{i=1}^{I}\sum_{c_j \in \mathcal{N}_i} \gamma_{ij}^T(x_j - z_{ij}) + \frac{d}{2}\sum_{i=1}^{I}\sum_{c_j \in \mathcal{N}_i}||x_j - z_{ij}||_2^2,$$

in which $\{\beta_{ij}\}$ and $\{\gamma_{ij}\}$ are Lagrange multipliers and d is a positive constant. At time t, the ADMM optimizes the augmented Lagrangian function as follows:

Step 1: Optimizing the local copies $\{x_i\}$:

$$\{x_i(t+1)\} = \arg\min_{\{x_i\}} \quad L_a\left(\{x_i\}, \{z_{ij}(t)\}, \{\beta_{ij}(t)\}, \{\gamma_{ij}(t)\}\right). \tag{8}$$

Notice that the objective function is separable, $x_i(t+1)$ can be updated as:

$$x_i(t+1) = \arg\min_{x_i} \quad (\frac{\lambda}{2}||F_i x_i - b_i||_2^2 + \frac{1}{I}1^T x_i)$$
$$+ \sum_{c_j \in \mathcal{N}_i} \beta_{ij}^T(t)x_i + \sum_{c_j \in \mathcal{N}_i} \gamma_{ji}^T(t)x_i \tag{9}$$
$$+ \frac{d}{2}\sum_{c_j \in \mathcal{N}_i}||x_i - z_{ij}(t)||_2^2 + \frac{d}{2}\sum_{c_j \in \mathcal{N}_i}||x_i - z_{ji}(t)||_2^2.$$

Step 2: Optimizing the Auxiliary Variable $\{z_{ij}\}$:

$$\{z_{ij}(t+1)\} = \arg\min_{\{z_{ij}\}} \quad L_a\left(\{x_i(t+1)\}, \{z_{ij}\}, \{\beta_{ij}(t)\}, \{\gamma_{ij}(t)\}\right). \tag{10}$$

Here the objective functions is also separable. Therefore:

$$z_{ij}(t+1) = \arg\min_{z_{ij}} \quad -\beta_{ij}^T(t)z_{ij} - \gamma_{ji}^T(t)z_{ij}$$
$$+ \frac{d}{2}||x_i(t+1) - z_{ij}||_2^2 + \frac{d}{2}||x_j(t+1) - z_{ij}||_2^2. \tag{11}$$

It has an explicit solution:

$$z_{ij}(t+1) = \frac{1}{2}\left(x_i(t+1) + x_j(t+1)\right) + \frac{1}{2d}\left(\beta_{ij}(t) + \gamma_{ij}(t)\right). \tag{12}$$

Step 3: Updating the Lagrange Multipliers $\{\beta_{ij}\}$ and $\{\gamma_{ij}\}$:

$$\beta_{ij}(t+1) = \beta_{ij}(t) + d\left(x_i(t+1) - z_{ij}(t+1)\right),$$
$$\gamma_{ij}(t+1) = \gamma_{ij}(t) + d\left(x_j(t+1) - z_{ij}(t+1)\right). \tag{13}$$

The updating rules of (9), (11), and (13) can be further simplified. Substituting (12) to (13) yields:

$$\beta_{ij}(t+1) = \beta_{ij}(t) + \frac{d}{2}\left(x_i(t+1) - x_j(t+1)\right) - \frac{1}{2}(\beta_{ij}(t) + \gamma_{ij}(t)),$$
$$\gamma_{ij}(t+1) = \gamma_{ij}(t) + \frac{d}{2}\left(x_j(t+1) - x_i(t+1)\right) - \frac{1}{2}(\beta_{ij}(t) + \gamma_{ij}(t)). \tag{14}$$

Since we often set $\beta_{ij}(0) = \gamma_{ij}(0) = \mathbf{0}$ where $\mathbf{0}$ denotes a $K \times 1$ all-zero vector $[0,0,...,0]^T$, (14) implies that $\beta_{ij}(t) = -\gamma_{ij}(t) = \gamma_{ji}(t)$. Then (12) becomes:

$$z_{ij}(t+1) = \frac{1}{2}\left(x_i(t+1) + x_j(t+1)\right). \tag{15}$$

and (13) becomes

$$\beta_{ij}(t+1) = \beta_{ij}(t) + \frac{d}{2}\left(x_i(t+1) - x_j(t+1)\right) = \gamma_{ji}(t+1),$$
$$\gamma_{ij}(t+1) = \gamma_{ij}(t) + \frac{d}{2}\left(x_j(t+1) - x_i(t+1)\right) = \beta_{ji}(t+1). \tag{16}$$

Summarizing the three sides of (16) and define a new Lagrangian multiplier $\alpha_i = \frac{2}{|\mathcal{N}_i|}\sum_{c_j \in \mathcal{N}_i} \beta_{ij} = \frac{2}{|\mathcal{N}_i|}\sum_{c_j \in \mathcal{N}_i} \gamma_{ji}$, the updating rule for α_i is:

$$\alpha_i(t+1) = \alpha_i(t) + dx_i(t+1) - \frac{d}{|\mathcal{N}_i|}\sum_{c_j \in \mathcal{N}_i} x_j(t+1). \tag{17}$$

Substituting (15) to (9), we have the updating rule for x_i:

$$x_i(t+1) = \arg\min_{x_i} \quad \sum_{i=1}^{I}(\frac{\lambda}{2}||\mathbf{F}_i x_i - \mathbf{b}_i||_2^2 + \frac{1}{I}\mathbf{1}^T x_i)$$
$$+|\mathcal{N}_i|\alpha_i^T(t)x_i + d\sum_{c_j \in \mathcal{N}_i}||x_i - \frac{1}{2}\left(x_i(t) + x_j(t)\right)||_2^2. \tag{18}$$

Iteratively solving (18) and updating (17) leads to the optimal solution of (4).

It should be noted that the problem we are discussing is indeed a special case of compressive sensing (Donoho et al., 2006). When the number of sensor nodes is smaller than the number of grid points, down-sampling is achieved via exploiting the sparse nature of the signal x. From this viewpoint, the proposed decentralized sparse signal recovery algorithm is also applicable to other compressive sensing problems, in which distributed sensor nodes hold parts of measurement matrices as well as measurement vectors, and collaboratively make decisions.

4.2 Algorithm outline

The collaborative sparse signal recovery algorithm is summarized as follows. The algorithm is fully decentralized, requiring only collaboration between neighboring cluster heads.

Algorithm: Collaborative environmental monitoring

Step 1: Initialization. At each sampling point, each cluster head collects position information and measurements from sensors within its cluster. Hence cluster head c_i knows the partial measurement matrix $\mathbf{F}_i$ and the partial measurement vector $\mathbf{b}_i$. The number of cluster heads I, the non-negative constant λ, and the positive constant d are also known.

Step 2: Communication. At iteration $t + 1$, each cluster head c_i broadcasts to its neighboring cluster heads to acquire intermediate decision vectors $x_j(t)$ of iteration t, $c_j \in \mathcal{N}_i$.

Step 3: Optimization. At iteration $t + 1$, each cluster head c_i updates its Lagrange multiplier $\alpha_i(t + 1)$ and decision vector $x_i(t + 1)$ according to (17) and (18).

Step 4: Iteration. Repeat Step 2 and Step 3 until convergence.

5. Performance analysis

In this section we will briefly discuss the impact of parameter settings on the performance of the algorithm, as well as the design choice of a hierarchical wireless sensor network. By performance we are mainly concerning: 1) quality of recovery, which includes the number of false alarms and the gap between the true and estimated amplitudes; and 2) convergence rate, which directly decides the communication burden of the cluster heads.

5.1 Parameter settings

The role of the non-negative weight λ in (3) has been extensively discussed in compressive sensing literature, such as (Donoho et al., 2006; Figueiredo et al., 2007). There is a constant $\lambda_{min} = 1/||F^T b||_\infty$, such that if $\lambda \leq \lambda_{min}$ the optimal solution is 0. When λ goes to infinity, the optimal solution has the minimum ℓ_1 norm among all points that satisfy $F^T(Fx - b = 0)$, if these points exist. Hence if b is noise-free and F is a full rank square matrix, then the optimal solution goes to the true signal. However large λ generally leads to a non-sparse solution under the existence of measurement noise.

In Step 1 of the collaborative environmental monitoring algorithm, we need to know I, the number of cluster heads. This procedure requires multi-hop communications if the cluster heads are not directly connected with each other. However, accurate knowledge of I is not necessary since it is the product λI that decides the optimal solution.

The proposed algorithm converges for any given positive constant d; however, the value of d influences the convergence rate, and thus the communication burden. It is possible to dynamically increase the value of d to infinity to improve the convergence rate during the iterative optimization process (Bertsekas and Tsitsiklis, 1997). Due to the extra burden of updating d, we simply choose d as a constant.

One of the most important advantages of the hierarchical infrastructure is its flexibility to different application scenarios. By setting $I = 1$, the infrastructure turns to be centralized; while with $I = L$ and sensors being cluster heads, the network is a fully decentralized one. This flexibility enables the network to adapt to different application scenarios.

5.2 To Collaborate or not to collaborate

Now we revaluate the order of the required communication load in a hierarchical network without any fusion center. First, at the data acquisition stage, cluster heads need to collect measurements from sensor nodes and construct the local measurement matrix, at communication cost on the order $\sim O(L)$. Second, at the optimization stage, one cluster head needs to transmit its decision vector to each neighboring cluster head at each iteration. The lengths of decision vectors are all K; the average number of neighboring cluster heads varies from $\sim O(1)$ (multi-hop communications) to $\sim O(I)$ (one-hop communications) depending on the communication range r_C of cluster heads. Denote the number of iterations as T, which is in general influenced by the choices of d and λ, as well as the topology of the network. Therefore the overall communication load ranges from $O(KIT)$ to $O(KI^2T)$.

For comparison, we also consider the communication load of a hierarchical network in the presence of fusion center. The communication load at the data acquisition stage is the same as before. Then each cluster head needs to transmit its own part of the measurement vector and the measurement matrix to the fusion center. Transmission of the measurement matrix is necessary since the wireless sensor network is often subject to node failure, node displacement, etc; hence the measurement matrix is generally a dynamic one. Suppose that sensor nodes are evenly divided into clusters; as a result, each cluster head has $1/I$ of the entire measurement matrix with KL entries and the entire measurement vector with L data. Depending on the communication range r_C of cluster heads, average communication load of sending one packet to the fusion center ranges from $O(1)$ (one-hop communications) to $O(I)$ (multi-hop communications). Thus, the overall communication load is from $O(KL)$ to $O(KIL)$.

The analysis provides us guideline on whether to use a fusion center or to implement decentralized collaborative information processing among cluster heads. When the number of sensor nodes within each cluster is large (large amount of data) and when the cluster heads are subject to multi-hop communications (limited communication range), the collaborative algorithm is superior in terms of energy efficiency. For a large-scale wireless sensor network, each cluster generally contains a large number of sensor nodes and the range of the sensing area is much larger than the communication range of cluster heads. Therefore, decentralized collaboration among cluster heads is preferred. In addition, collaborative information processing offers the benefits of robustness to node failure, obviation of multi-hop routing, and alleviated level of congestion.

6. Simulation results

Let us consider a 100×100 sensing area with length from 0 to 100 and width from 0 to 100. The area is divided into 100 squares with 121 grid points. Sensor nodes are uniformly randomly deployed in the sensing area. Two sources of events, one with amplitude 1 and another with amplitude 0.5, occur at grid points $(20, 80)$ and $(50, 50)$ respectively. We assume the influence function to be $f_{kl} = \exp(-d_{kl}^2/\sigma^2)$, with $\sigma = 20$. Two parameters of the decentralized algorithm are set to $\lambda = 100$ and $d = 1$.

First, we consider 100 sensor nodes, which are divided into 5 clusters with 20 sensor nodes in each cluster. The cluster heads are bi-directionally connected to each other by properly setting the communication range r_C. As an ideal case, the measurements are assumed to be noise-free. The convergence property of one cluster head is depicted in Figure 3. The decision variables corresponding to the two events converge to the optimal values, while the amplitudes of other grid points converge to 0. Due to the consensus constraints, decision vectors of different cluster heads converge to the same solution.

Next, we study the influence of the number of cluster heads I on the convergence time. By convergence time we mean the minimum iteration number with which the differences between all decision variables and their optimal values are within 0.01. According to Figure 4, it is not surprising that the fully centralized infrastructure ($I = 1$) achieves the best convergence rate while the fully decentralized one ($I = 100$) converges slowly. Figure. 4 indicates that the convergence time is $\sim O(I^2)$ for the decentralized algorithm.

Further, to compare the communication load between the two schemes, we assume all sensor nodes and cluster heads have a common communication range $r_C = 10$. In the centralized scheme, one sensor node is chosen as the fusion center. In the decentralized scheme, each cluster head is supposed to have at least one neighboring cluster head, since

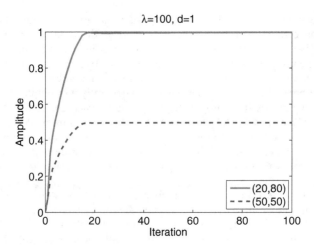

Fig. 3. Convergence of the proposed algorithm.

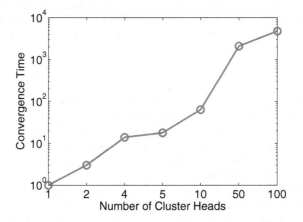

Fig. 4. Number of cluster heads I versus convergence time.

the connectivity of the sub-network of cluster heads generally cannot be satisfied when the number of cluster heads is small. The communication loads are depicted in Figure. 5. It is shown that when the cluster heads collect a large amount of data, collaborative optimization is energy-efficient. When the number of cluster heads increases, the communication load for consensus optimization dominates, leading to poor efficiency. This fact suggests us to properly select the cluster size and number of clusters in order to be energy efficient. It also points out an important but challenging research topic for future work, namely, improving the energy-efficiency of hierarchical networks via accelerating convergence rate of the decentralized collaborative algorithm.

Finally we simply discuss the compressive ratio of the proposed algorithm, namely, the ratio of the number of sensor nodes versus the number of grid points. We maintain the parameter settings in the previous simulation; the number of cluster heads is not necessary since any

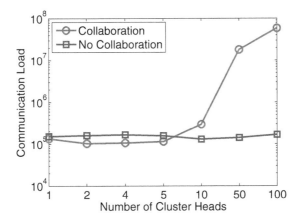

Fig. 5. Number of cluster heads I versus communication load.

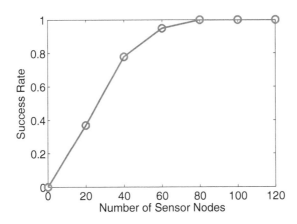

Fig. 6. Number of sensor nodes L versus the success rate of recovery.

settings will lead to global convergence. The relationship between the compressive ratio and the probability of successful recovery is shown in Figure 6. The simulation is repeated for 100 times with randomly deployed sensor nodes for each time. When the number of sensor nodes is larger than nearly half of the number of grid points, the recovery is successful with high probability.

7. Conclusion

This chapter discusses the design of hierarchical wireless sensor networks for environmental monitoring applications. Specifically, we focus on a generalized event detection model which is able to discover sparse events based on sensory data. Both positions and amplitudes of the events can be recovered from a convex program. Then we elaborate on an optimal decentralized algorithm which requires no fusion center but only collaboration of neighboring cluster heads. Through theoretical analysis and simulation experiments, we suggest when the cluster heads need to collaborate and when not; this provides a design guideline for hierarchical wireless sensor networks.

8. Acknowledgement

Qing Ling is supported in part by National Nature Science Foundation under grant 61004137 and the Fundamental Research Funds for the Central Universities under grant WK2100100007.

9. References

I. Akyildiz, W. Su, Y. Sankarasubramaniam, and E. Cayirci, "Wireless sensor networks: a survey," Computer Networks, vol. 38, pp. 393–422, 2002

J. Yick, B. Mukherjee, and D. Ghosal, "Wireless sensor network survey," Computer Networks, vol. 52, pp. 2292–2330, 2008

D. Estrin, R. Govindan, J. Heidemann, and S. Kumar, "Next century challenges: scalable coordination in sensor networks," In: Proceedings of MOBICOM, 1999

D. Estrin, D. Culler, K. Pister, and G. Sukhatme, "Connecting the physical world with pervasive networks," IEEE Pervasive Computing, vol. 1, pp. 59–69, 2002

G. Simon, M. Maroti, A. Ledeczi, G. Balogh, B. Kusy, A. Nadas, G. Pap, J. Sallai, and K. Frampton, "Sensor network-based countersniper system," SENSYS 2004

T. He, S. Krishnamurthy, L. Luo, T. Yan, L. Gu, R. Stoleru, G. Zhou, Q. Cao, P. Vicaire, J. Stankovic, T. Abdelzaher, J. Hui, and B. Krogh, "VigilNet: an integrated sensor network system for energy-efficient surveillance," ACM Transactions on Sensor Networks, vol. 2, pp. 1–38, 2006

K. Langendoen, A. Baggio, and O. Visser, "Murphy loves potatoes: experiences from a pilot sensor network deployment in precision agriculture," In: Proceedings of IPDPS, 2006

T. Wark, P. Corke, P. Sikka, L. Klingbeil, Y. Guo, C. Crossman, P. Valencia, D. Swain, and G. Bishop-Hurley, "Transforming agriculture through pervasive wireless sensor networks," IEEE Pervasive Computing, vol. 6, pp.. 50–57, 2007

V. Gungor and G. Hancke, "Industrial wireless sensor networks: challenges, design principles, and technical approaches," IEEE Transactions on Industrial Electronics, vol. 56, pp. 4258–4265, 2009

M. Li and Y. Liu, "Underground structure monitoring with wireless sensor networks," In: Proceedings of IPSN, 2007

Q. Ling, Z. Tian, Y. Yin, and Y. Li, "Localized structural health monitoring using energy-efficient wireless sensor networks," IEEE Sensors Journal, vol. 9, pp. 1596–1604, 2009

P. Zhang, C. Sadler, S. Lyon, and M. Martonosi, "Hardware design experiences in ZebraNet," In: Proceedings of SENSYS, 2004

P. Corke, T. Wark, R. Jurdak, W. Hu, P. Valencia, and D. Moore, "Environmental wireless sensor networks," Proceedings of the IEEE, vol. 98, pp. 1903–1917, 2010

B. Sadler, "Fundamentals of energy-constrained sensor network systems," IEEE Aerospace and Electronic Systems Magazine, vol. 20, pp. 17–35, 2005

G. Werner-Allen, J. Johnson, M. Ruiz, J. Lees, and M. Welsh, "Monitoring volcanic eruptions with a wireless sensor network," In: Proceedings of EWSN, 2005

G. Werner-Allen, K. Lorincz, M. Welsh, O. Marcillo, J. Johnson, M. Ruiz, and J. Lees, "Deploying a wireless sensor network on an active volcano," IEEE Internet Computing, vol. 10, pp. 18–25, 2006

Y. Liu, Y. He, M. L, J. Wang, K. Liu, L. Mo, W. Dong, Z. Yang, M. Xi, J. Zhao, and X. Li, "Does wireless sensor network scale? a measurement study on GreenOrbs," In: Proceedings of INFOCOM, 2011

Q. Ling and Z. Tian, "Decentralized sparse signal recovery for compressive sleeping wireless sensor networks,ąś IEEE Transactions on Signal Processing, vol. 58, pp. 3816–3827, 2010

A. Arora, R. Ramnath, E. Ertin, P. Sinha, S. Bapat, V. Naik, V. Kulathumani, H. Zhang, H. Cao, M. Sridharan, S. Kumar, N. Seddon, C. Anderson, T. Herman, N. Trivedi, C. Zhang, M. Nesterenko, R. Shah, S. Kulkarni, M. Aramugam, L. Wang, M. Gouda, Y. Choi, D. Culler, P. Dutta, C. Sharp, G. Tolle, M. Grimmer, B. Ferriera, and K. Parker, "ExScal: elements of an extreme scale wireless sensor network," In: Proceedings of RTCSA, 2005

P. Dutta, J. Hui, J. Jeong, S. Kim, C. Sharp, J. Taneja, G. Tolle, K. Whitehouse, and D. Culler, "Trio: enabling sustainable and scalable outdoor wireless sensor network deployment," In: Proceedings of IPSN, 2006

G. Barrenetxea, F. Ingelrest, G. Schaefer, and M. Vtterli, "The hitchhiker's guide to successful wireless sensor network deployment," In: Proceedings of SENSYS, 2008

W. Heinzelman, A. Chandrakasan, and H. Balakrishnan, "An application-specific protocol architecture for wireless microsensor networks," IEEE Transactions on Wireless Communications, vol. 1, pp. 660–670, 2002

M. Rabbat and R. Nowak, "Distributed optimization in sensor networks," In: Proceedings of IPSN, 2004

S. Aldosari, J. Moura, "Fusion in sensor networks with communication constraints," In: Proceedings of IPSN, 2004

D. Bertsekas and J. Tsitsiklis, *Parallel and Distributed Computation: Numerical Methods*, Second Edition, Athena Scientific, 1997

D. Donoho, M. Elad, V. Temlyakov, "Stable recovery of sparse overcomplete representations in the presense of noise," IEEE Transactions on Information Theory, vol. 52, pp. 6–18, 2006

R. Szewcszyk, A. Mainwaring, J. Polastre, J. Anderson, and D. Culler, "Lessons from a sensor network expedition," In: Proceedings of EWSN, 2004

G. Tolle, J. Polastre, R. Szewczyk, D. Culler, N. Turner, K. Tu, S. Burgess, T. Dawson, P. Buonadonna, D. Gay, and W. Hong, "A macroscope in the redwoods," In: Proceedings of SENSYS, 2005

L. Selavo, A. Wood, Q. Cao, T. Sookoor, H. Liu, A. Srinivasan, Y. Wu, W. Kang, J. Stankovic, D. Young, and J. Porter, "LUSTER: wireless sensor network for environmental research," In: Proceedings of SENSYS, 2007

K. Martinez, J. Hart, and R. Ong, "Environmental sensor networks," Computer, vol. 37, pp. 50–56, 2004

J. Bezdek, S. Rajasegarar, M. Moshtaghi, C. Leckie, M. Palaniswami, and T. Havens, "Anomaly detection in environmental monitoring networks," IEEE Computational Intelligence Magazine, vol. 6 pp. 52–58, 2011

C. Alippi, R. Camplani, G. Galperti, and M. Roveri, "A robust, adaptive, solar-powered WSN framework for aquatic environmental monitoring," IEEE Sensors Journal, vol. 11, pp. 45–55, 2011

J. Bazerque and G. Giannakis, "Distributed spectrum sensing for cognitive radio networks by exploiting sparsity," IEEE Transactions on Signal Processing, vol. 58, pp. 1847–1862, 2010

S. Kim, K. Koh, M. Lustig, S. Boyd, and D. Gorinevsky, "An interior-point method for large-scale ℓ-1 regularized least squares," IEEE Journal of Selected Topics in Signal Processing, vol. 1, pp. 606–617, 2007

M. Figueiredo, R. Nowak, and S. Wright, "Gradient projection for sparse reconstruction: application to compressed sensing and other inverse problems," IEEE Journal of Selected Topics in Signal Processing, vol. 1, pp. 586–597, 2007

H. Zhu, A. Cano, and G. Giannakis, "Consensus-based distributed MIMO decoding using semidefinite relaxation," In: Proceedings of CAMSAP, 2007

Permissions

The contributors of this book come from diverse backgrounds, making this book a truly international effort. This book will bring forth new frontiers with its revolutionizing research information and detailed analysis of the nascent developments around the world.

We would like to thank Dr. E.O. Ekundayo, for lending his expertise to make the book truly unique. He has played a crucial role in the development of this book. Without his invaluable contribution this book wouldn't have been possible. He has made vital efforts to compile up to date information on the varied aspects of this subject to make this book a valuable addition to the collection of many professionals and students.

This book was conceptualized with the vision of imparting up-to-date information and advanced data in this field. To ensure the same, a matchless editorial board was set up. Every individual on the board went through rigorous rounds of assessment to prove their worth. After which they invested a large part of their time researching and compiling the most relevant data for our readers. Conferences and sessions were held from time to time between the editorial board and the contributing authors to present the data in the most comprehensible form. The editorial team has worked tirelessly to provide valuable and valid information to help people across the globe.

Every chapter published in this book has been scrutinized by our experts. Their significance has been extensively debated. The topics covered herein carry significant findings which will fuel the growth of the discipline. They may even be implemented as practical applications or may be referred to as a beginning point for another development. Chapters in this book were first published by InTech; hereby published with permission under the Creative Commons Attribution License or equivalent.

The editorial board has been involved in producing this book since its inception. They have spent rigorous hours researching and exploring the diverse topics which have resulted in the successful publishing of this book. They have passed on their knowledge of decades through this book. To expedite this challenging task, the publisher supported the team at every step. A small team of assistant editors was also appointed to further simplify the editing procedure and attain best results for the readers.

Our editorial team has been hand-picked from every corner of the world. Their multi-ethnicity adds dynamic inputs to the discussions which result in innovative outcomes. These outcomes are then further discussed with the researchers and contributors who give their valuable feedback and opinion regarding the same. The feedback is then collaborated with the researches and they are edited in a comprehensive manner to aid the understanding of the subject.

Apart from the editorial board, the designing team has also invested a significant amount of their time in understanding the subject and creating the most relevant covers. They scrutinized every image to scout for the most suitable representation of the subject and create an appropriate cover for the book.

The publishing team has been involved in this book since its early stages. They were actively engaged in every process, be it collecting the data, connecting with the contributors or procuring relevant information. The team has been an ardent support to the editorial, designing and production team. Their endless efforts to recruit the best for this project, has resulted in the accomplishment of this book. They are a veteran in the field of academics and their pool of knowledge is as vast as their experience in printing. Their expertise and guidance has proved useful at every step. Their uncompromising quality standards have made this book an exceptional effort. Their encouragement from time to time has been an inspiration for everyone.

The publisher and the editorial board hope that this book will prove to be a valuable piece of knowledge for researchers, students, practitioners and scholars across the globe.

List of Contributors

Wei Song and Danke Xu
State Key Laboratory of Analytical Chemistry for Life Science, School of Chemistry and Chemical Engineering, Nanjing University, China

Si Wei and Hong-Xia Yu
State Key Laboratory of Pollution Control and Resource Reuse, School of the Environment, Nanjing University, China

Maika Vuki
College of Natural and Applied Sciences, University of Guam, Mangilao, Guam, USA

Elisa Benetti, Chiara Taddia and Gianluca Mazzini
Lepida SpA, Viale A. Moro 64, 40127 Bologna, Italy

Raffaele Giordano, Giuseppe Passarella and Emanuele Barca
Water Research Institute - National Research Council, Bari, Italy

Philippe Gourbesville
Nice Sophia Antipolis University / Polytech Nice Sophia, France

Akihide Utani
Tokyo City University, Japan

Ittipong Khemapech
University of the Thai Chamber of Commerce, Thailand

Bernd Resch, Rex Britter, Christine Outram, Xiaoji Chen and Carlo Ratti
Massachusetts Institute of Technology, USA

Qing Ling and Gang Wu
Department of Automation, University of Science and Technology of China, China

Zhi Tian
Department of Electrical and Computer Engineering, Michigan Technological University, USA

Printed in the USA
CPSIA information can be obtained
at www.ICGtesting.com
JSHW011348221024
72173JS00003B/238